Nanoparticle Engineering for Chemical-Mechanical Planarization

Fabrication of Next-Generation Nanodevices

Nanoparticle Engineering for Chemical-Mechanical Planarization

Fabrication of Next-Generation Nanodevices

Ungyu Paik • Jea-Gun Park

CRC Press
Taylor & Francis Group
6000 Broken Sound Parkway NW, Suite 300
Boca Raton, FL 33487-2742

First issued in paperback 2019

ISBN-13: 978-0-367-44606-2 (pbk)
ISBN-13: 978-1-4200-5911-3 (hbk)

Visit the Taylor & Francis Web site at
http://www.taylorandfrancis.com

and the CRC Press Web site at
http://www.crcpress.com

Contents

Preface

As a result of a ubiquitous society, displays for electronic appliances such as mobile computers and mobile phones demand high efficiency. High efficiency electronics were realized with the advent of nanoscale devices. Especially for nanoscale devices, the importance of the chemical–mechanical planarization (CMP) process emerged to achieve the integration and multilevel interconnections. The CMP process supports increasing capacity.

CMP technology became absolutely necessary for production processes of the next generation of semiconductor. Pursuing global planarization, planarization using CMP could not be against the trend. Therefore, it is used for production of device makers, which is dielectric CMP, shallow trench isolation CMP, and metal CMP. Chapters 2 through 4, respectively, discuss these processes. The nanotopography of the surface of silicon wafers has recently become an important issue because it may seriously affect the post-CMP uniformity of thickness variation of dielectrics. For this reason, Chapter 5 explains the importance of nanotopography.

However, CMP processing has faced a new aspect: design rules below 50 nm. Because new structures and new materials are used for improving the performance of devices, the existing CMP slurry is facing its limits. Chapter 6 provides novel CMP slurry for application to memory devices beyond 50 nm technology.

The authors are especially grateful to Allison Shatkin and Amy Blalock at CRC Press for their valuable guidance.

Ungyu Paik

Jea-Gun Park

The Authors

Ungyu Paik is professor and director of the Division of Materials Science Engineering at Hanyang University, Korea. He is head of the Research Center for Converging Technology in Advanced Gas Turbine System and director of the Global Research Laboratory for Nano Device Processing Laboratory. He is also a technology counselor for the Materials Laboratory of Samsung SDI Corporate Research & Development Center and Hynix semiconductor. Paik received a PhD from the Department of Ceramic Engineering at Clemson University, Clemson, South Carolina, in 1991. Prior to joining Hanyang University, he was an associate professor at the Department of Materials Engineering, Changwon National University, Korea, from 1992 to 1999 and worked as a guest researcher at National Institute of Standards and Technology, Gaithersburg, Maryland, from 1995 to 1996.

Paik's research interests are in the control of the interparticle force of nanoceramic particles and development of nanoparticle patterning technology. His newly expanded topics are focused on dispersion stability of carbon nanotubes (CNTs), quantum dot in non-volatile memory, phosphor in plasma display panel (PDP), color filter in liquid-crystal display (LCD), and so on. His technology transfer experience includes: the development of fabrication technology of ultra-thin (1.2 μm) dielectric sheet in multilayered ceramic capacitor (MLCC) and technology transfer to Samsung Electro-Mechanics; the world's first design of dispersion technology of aqueous graphite for the application of lithium ion battery anodes and technology transfer for Samsung SDI Co.; the technology development of CMP slurry for interlayer deposition and the technology transfer for Technosemichem Co.; and the initiation of technology development of high performance CMP slurry for shallow trench isolation for KC Tech Co. He has published more than 100 technical papers in the scientific literature and holds 38 patents. Also, he gave several plenary and invited talks at international conferences, including SEMICON Korea 2003, the 54th Pacific Coast Regional & Basic Science Division Meeting of the American Ceramic Society, and the 1st International Congress on Ceramics of the American Ceramic Society. He is a recognized world leader in dispersion technology.

Jea-Gun Park

Jea-Gun Park is professor of the Division of Electrical & Computer Engineering, director of the Department of Nanoscale Semiconductor Engineering, director of the Industry-University Cooperation Foundation, and dean of the University Research at Hanyang University, Korea. He has been interested in defect engineering in semiconductor materials and nano-scale device development since he received a Ph.D. at the Department of Material Science & Engineering, North Carolina State University, U.S.A. His interests in defect engineering continued during a 17-year career at the Samsung Electronics Co., Semiconductor Division (1985-2001) and broadened beyond defect engineering to the development of "Pure Silicon Wafer" (free of agglomerated defects in CZ Si), which has been used as a standard wafer for DRAM devices. Since coming to Hanyang University in 1999, his research interests have developed "Super Silicon Wafer" (pure silicon wafer containing extremely proximity gettering effect), which has been used as a standard wafer for flash memory devices. In 2001, Prof. Jea-Gun Park and his team developed "300mm Super Silicon Wafer," which is immune to metallic ions contamination in the semiconductor process line for the first time in the world and standardized as an international standard silicon substrate for sub-100nm C-MOSFET through his technology transfer to Japanese silicon wafer manufactures, Shinetsu Handotai, Mitsubishi Materials, and Sumitomo Metal. Nowadays, it is being applied to Samsung, Hynix, and Toshiba for mass production. In addition, he developed nano SOI process technology for giga-hertz speed CPU or MPU consisting of SOI C-MOSFET, and high mobility strained-Si C-MOSFET. Furthermore, one of his main research areas is nano-CMP slurries. In particular, he developed fumed silica slurry for inter-layer dielectric process in 2001 and transferred it to TECHNOCHEM (Korean company). In addition, he also developed nano-ceria slurry for shallow trench isolation process in 2003 and transferred it to KCTECH (Korean company). Currently, he is doing the research & development of the process technologies and slurries for high performance Cu/low-k CMP, poly-Si CMP, and GST CMP. Moreover, he is also developing flexible & transparent Si OLED (Organic Light Emission Display) as well as flexible & transparent 30nm OBLED (Organic Bistable OLED: Memory Transistor + OLED) for the application of high resolution micro display. He is focusing on the research and development for the process, device, and circuit design of PoRAM (Polymer Memory) which is promising next generation transistor of tera-bit non-volatile memory. Currently, professor Jea-Gun Park is taking a role as a managing director of the technology development program for the Next Generation of Terabit Non-volatile Memory sponsored by Korea MKE (Ministry of Knowledge Economy), a head of High-speed/Cap-less Memory Research Sponsored by Korea MEST (Ministry of Education, Science and Technology), a chairman of VLSI symposium in Electrochemical Society, ASIA representative of SIWEDS (Silicon Wafering Engineering & Defect Science) partly sponsored by NSF of the USA, and a steering committee member of Korea advanced national nano-FAB. As of Oct. 20, 2008, he had published 172 papers on an international journal (SCI). In addition, he has submitted 147 patents and 84 patents have been registered. Prof. Jea-Gun Park has also presented 227 talks.

1

Overview of CMP Technology

1.1 Motivation and Background

Since Bardeen, Brattain, and Shockley of Bell Laboratories invented the transistor in the 1940s, semiconductor integrated circuit (IC) technology has been remarkably developed. The improvement of operation capacity and speed, thanks to the development of semiconductor technology, is playing a key role in the rapid progress of current scientific technologies. In the flow of rapid progress, it was required for semiconductors to possess super high speed, capacity, and performance, and as a result, integration of the transistor is increasing. Since the beginning of the 21st century, the design rule of semiconductors was set below 100 nm. Samsung Electronics developed 64 Gb NAND flash memory using 30 nm technology in 2007, Hynix developed 2G DRAM2 using 66 nm technology, and Intel developed the Core™2 Quad Processor using 65 nm technology. Although these remarkable developments pursue a miniature through vertical high integrated circuit, it is also possible to obtain super high speed and capacity through a horizontal, high integrated circuit (multiple metal lines). In multilayer metal lines process technology, it is difficult to focus and impossible to form minute structures when the dielectric layer and metal line have rugged surfaces. Therefore, planarization processing was necessary to ensure lithographic depth of focus (DOF), which was considered to be the most important factor. A variety of planarization methods of high degree were indispensable in using a new material and transformation from two-dimensional flat structures to three-dimensional multilayer structures into a wide and high integrated ultra large scale integrated (ULSI) circuit device with a diameter from 200 mm to 300 mm of silicon wafer. By using preexisting borophosphosilicate glass (BPSG) deposition and planarization methods such as reflow, spin on glass (SOG), and spin coating, it may soften rugged surfaces to some degree during dielectric layering (Table 1.1). However, problems such as aggravation of formation, position precision, rugged surface device according to multilayer metal line, and three-dimensional structures would occur (Figure 1.1).

TABLE 1.1
Types and Features of Existing Planarization

Types	Examples of Method	Features
Etching	Sputter, RIE, plasma etching	Easy process, difficult to control etching
Deposition	Bias sputter, Bias ECR, plasma chemical vapor deposition (CVD), RF plasma CVD	Damage concern, too much dust
Reflow floating	Reflow, SOG (spin on glass)	Easy to utilize, discontinuity, instability, establishment of migration
Selective growth	Selective CVD, selective epitaxial growth	Possible to fill only the required part (hole), low selective growth control, instability

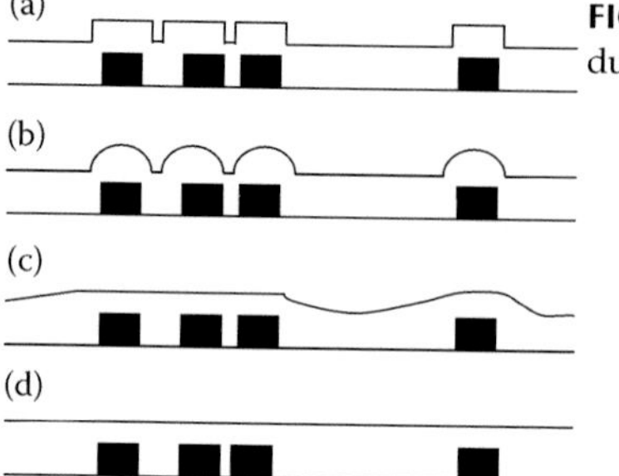

FIGURE 1.1 Formation of planarization of semiconductor metal line.

To solve this problem, Kaufmann, from IBM, developed in the 1980s chemical–mechanical planarization (CMP), a new global planarization technique that combines mechanical polishing with chemical polishing into a manufacturing process. This was the origin of CMP process in the modern sense (Figure 1.2). CMP chemically or mechanically polishes the semiconductor's surface for planarization. The chemical action affects productivity and polishing selectivity rate, whereas the mechanical action contributes to the smoothness of the surface. As in Figure 1.1c, the CMP process eliminated the rugged upper part of a surface in an orderly manner, regardless of the low area condition. CMP made global planarization possible without unevenness and solved the problem of existing techniques that caused deteriorated layers to form and lowered shape precision. For this reason, CMP technology became absolutely necessary for the production process for the next generation of semiconductors and it is used for production of device makers. Internationally, many businesses and lab organizations are actively making progress with research into a new process technology.

This CMP propagation and active research progress may well be the first departure of shallow trench isolation (STI) CMP. STI CMP was introduced

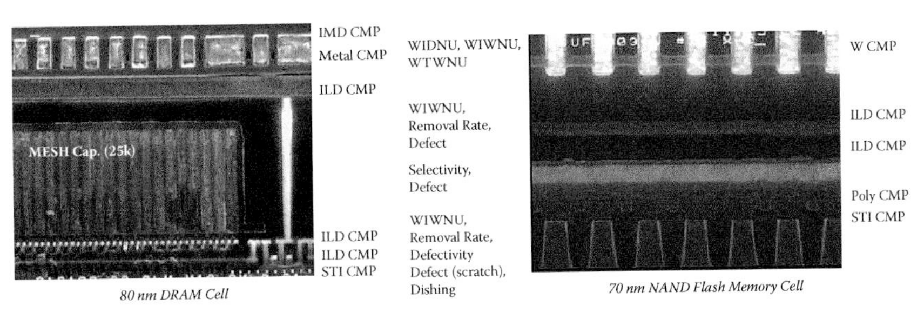

FIGURE 1.2 Global planarization by CMP.

by replacing local oxidation of silicon (LOCOS) with STI by the splitting method for each transistor. For the process of complementary metal-oxide semiconductor (CMOS), each transistor used LOCOS because separation by electricity was needed to eliminate a short. Unfortunately, this LOCOS process caused a severe problem with device integration because its design rule entered under 0.5 μm, making a sharp edge called bird's beak. To solve this problem, STI led to an increase in a very tiny active area and in device packing density. Despite the advantages of STI, it cannot be formed without the CMP process because to form STI, polishing must be stopped by eliminating gap filling oxide film at the Si_3N_4 layer. This process is only possible through STI CMP technology. Currently, manufactured semiconductor devices on the basis of CMOS are produced applying STI CMP.

1.2 The Key Factors of CMP Process

The characteristics of CMP are material removal rate (MRR), thickness uniformity, and surface quality, and they are directly related to device characteristics and productivity. These characteristics are determined by each factor as per Figure 1.3.

1.2.1 CMP Polishing Machines

Whereas wafer polishing machines polish dozens of micrometers (μm), CMP polishing machines polish 0.5 ~ 1.0 μm of target film. Uniformity is extremely important after polishing. The features of CMP polishing machines include automation, high precision, reproducibility, and control of process parameter. Polishing machines are largely divided into rotary type, orbital type, linear type, and fixed table type according to the movement of the wafer carrier and table (Table 1.2).

A schematic diagram of the Mirra polisher, currently used in a device maker, is shown in Figure 1.4. In the cassette load, the sensor perceives the

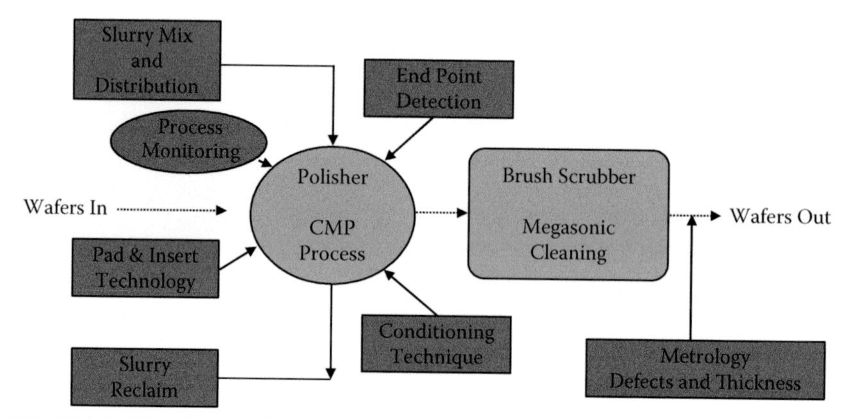

FIGURE 1.3 The elements (factors) of CMP process.

TABLE 1.2
Motion of Wafer Carrier and Table by CMP Tool Type

	Motion	
CMP Tool Type	Wafer Carrier	Table (Platen)
Rotary	Rotate	Rotate
Orbital	Rotate	Orbital path
Linear	Linear path	Linear path
Fixed table	Orbital path	Stationary

wafer, and the FABS robot moves to the transfer station. CMP is transferred to the head clean load unload (HCLU) where CMP is actually formed after a long robot arm absorbs the wafer by using a void space. The FAB robot arm transfers the finished CMP processed wafer to the input station for cleaning in the HCLU.

1.2.2 Slurry for CMP

As mentioned earlier, the core of CMP technology is slurry for CMP based on nanotechnology, even though the importance of CMP technology was highlighted before the introduction of a nano process for semiconductor processing. Slurry is composed of water, polishing particle, alkali, inorganic salts, and organic compound, and specific slurry is manufactured through appropriate selection of the components.

Some of the most important characteristics of slurry are equal dispersion of polished particle caused scratch on the surface of wafers, minimizing of metal current, optional polishing characteristics, viscosity, and

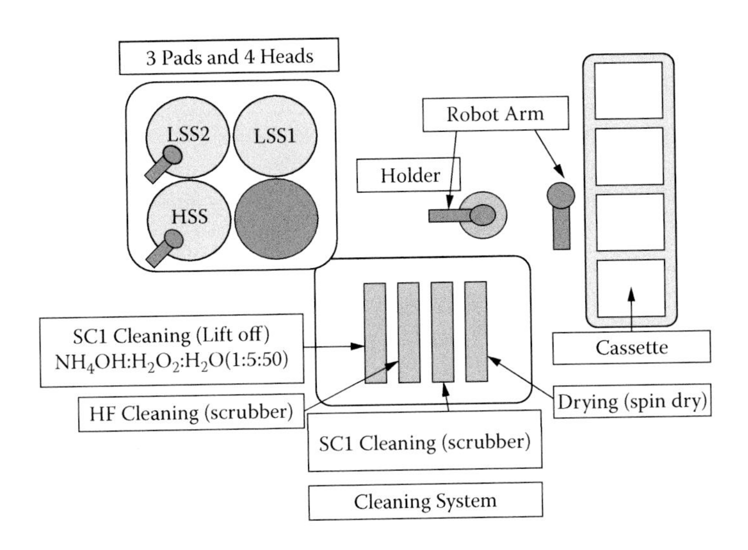

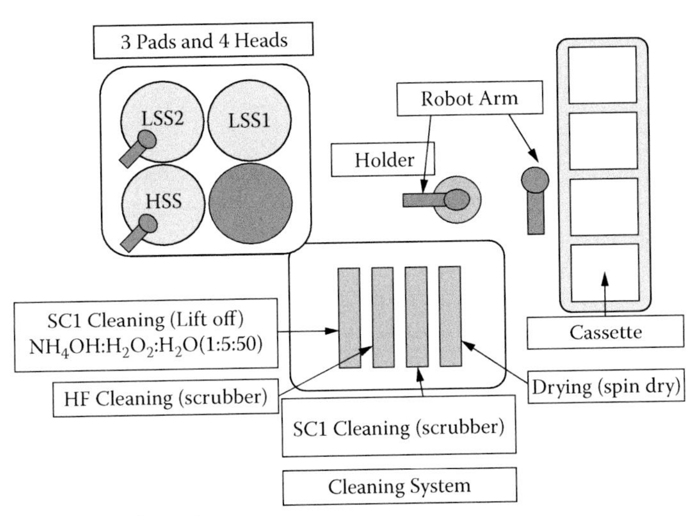

FIGURE 1.4 Mirra schematic.

storage safety that cause polysilicon abnormal resistance and metal line short. Commonly used solvent is highly pure, from which most of the impurities such as ion, small particles, microorganisms, and organic substance are eliminated, with a 5 ~ 18 MΩcm resistance rate. The appropriate slurry is applied according to objects to be polished. Table 1.3 summarizes the slurry required for various processes.

TABLE 1.3
CMP Slurry Type

Polishing Objects		Purpose of Application	Slurry Types
Interlayer dielectric	SiO_2 FSG BPSG Low-k	Planarization, trend favoring introduction of processing below 0.35 μm	Fumed silica is mainly used; tendency of switching to ceria
STI (shallow trench isolation)		Intra-semiconductor devices separation layer elimination; applied below 0.25 μm (128, 256M)	Ceria, fumed/colloidal silica
Polysilicon		Formation of trench-shaped capacitor, polysilicon is eliminated	Fumed/colloidal silica
Metal	W	Wire formation (W is eliminated)	Fumed silica is mainly used; alumina development phase
	Al	Contact plug formation (Al elimination)	Fumed silica, alumina
	Cu	Wiring and plug simultaneous formation (dual damascene: Cu is eliminated)	Fumed/colloidal silica, alumina, MnO_2, and others

1.2.3 Pad

Generally, polymer of polyurethane type is used for pads. Uniform surface roughness and porosity of pad influence the characteristics of WIWNU (within-wafer non-uniformity), WTWNU (wafer-to-wafer non-uniformity), and LTLNU (lot-to-lot non-uniformity). For this reason, chemical technology durability, hydrophilic, and viscoelastic features differ according to each required CMP process condition. Table 1.4 shows currently used pads types and the CMP process to which they are applied.

1.2.4 Slurry Supply Equipment and Filtering Equipment

The CMP process has a higher possibility of defects than other processes because it uses abrasive in slurry. It especially causes scratches; therefore, controlling the defects is important. To repress scratches attributed to slurry, filter is generally placed at the supply system, circulation loop, and point of use (POU). These factors can be mixed diversely according to the polishing machine's structure or processing condition selection. However, other materials are also influenced because of the correlation when a factor

TABLE 1.4

Typical Applications for Different Pad Types

	Type 1	Type 2	Type 3	Type 4
Structure	Felt fibers impregnated with polymeric binder	Porous film coated on a supporting substance	Microporous polymer sheet	Nonporous polymer sheet with surface macrotexture
Pad examples	Pellon™, Suba™	Polytex™, Surfin™, UR100™, WWP3000™	IC1000™, IC1010™, IC1400™, FX9™, MH™	OXP3000™, IC2000™
Typical applications	Si stock polish, tungsten CMP	Si final polish, tungsten CMP, post-CMP buff	Si stock, ILD CMP, STI, metal damascene CMP	ILD CMP, STI, metal dual damascene

changes. Therefore, each CMP process should be controlled appropriately because polishing target film and processing can be changed.

All CMP processes applied to semiconductor manufacturing processes, including STI CMP, are formed around the CMP machine. The surface of the wafer and a pad are contacted by pressure of its own load of a head part. At this time, a pad attached to the polishing table makes a simple rotary movement, and a head makes a rotary movement and shaking movement at the same time (Figure 1.5). The wafer exerts a regular pressure on the polishing table. Consumables are liquid slurry, a pad, and a cleaner and others conformable to each target substance. At this point, abrasive of the inside of slurry and the wafer device flow into interface

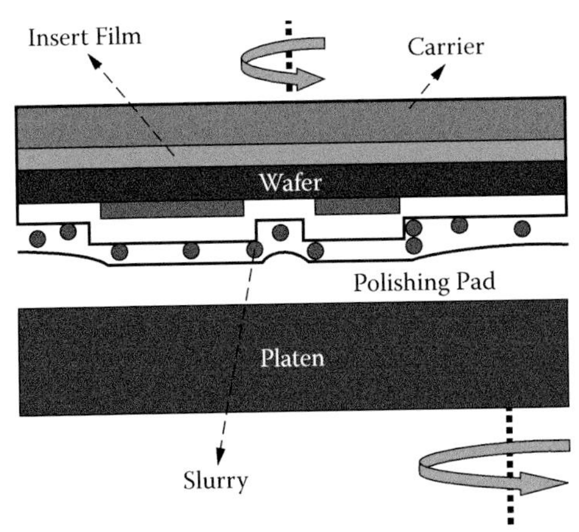

FIGURE 1.5 (See color insert) CMP process of manufacturing.

space to make contact at the overhang. Pressure is concentrated at this protuberant part. Therefore, it has a relatively higher speed of surface elimination than other parts. Also, the protuberant part is equally eliminated over the surface while processing is in progress.

Except for STI CMP, 50 nm processing technology CMP can be divided into interlayer dielectric (ILD) CMP and metal CMP. Chapters 2 and 6, respectively, will discuss these two processes. The 50 nm CMP processing rate rapidly augments through multiple metal lines structure for high integrated circuits and is recognized as a necessary process for formation of various detail patterns. The existing CMP slurry is becoming limited to future applications because the device design rule is going below the 50 nm level, and new materials and structures are emerging. In addition, structures that can be embodied only through CMP processing are appearing.

In the case of dynamic random access memory (DRAM), the top electrode used in capacitor for a device's high speed raises the necessity of noble metals like ruthenium (Ru), platinum (Pt), and iridium (Ir), which have low electric resistance and are mechanically and thermally stable. These noble metals are also chemically very stable and it is not easy to form capacitors by the etch back process. That is why noble metal CMP is compulsory. However, Ru is divided during the CMP process as a consequence of poor adhesion of leakage of cap oxide, grain growth of Ru, and cap oxide. To protect this phenomenon, the application of new functional slurry is essential.

NAND flash memory started to apply floating gate to increase the capacity of a device from 65 nano processing technology. After STI CMP processing in the gate formation area, silicon nitride is stripped. After Si is placed between the device manufacture areas using self-aligned poly (SAP) method, polysilicon floating gate is formed through CMP process. The poly gate isolation process using CMP raises many problems in applying the existing CMP slurry because of the soft characteristic of poly Si and polymeric reaction, despite simple processes like planarization after deposition.

Manufacturing of smaller devices necessitates the introduction of new materials and processes. The role of CMP is expanding and its importance is also being augmented. As a result, slurry production companies and laboratories are actively processing developments and researching consumable and optimized processing.

2

Interlayer Dielectric CMP

2.1 Interlayer Dielectric (ILD) CMP Process

In the deposition of the interlayer dielectric (ILD) film to break off relations wiring and wiring, step height is formed because the deposition aspect becomes different along the shape of the lower part pattern. Without removing induced step height in the wiring process, the limit of exposure is caused passing over the depth of focus (DOF) margin during the lithography process. Therefore, the global planarization process is essential after each layer is insulated, and this CMP process is designated as ILD CMP or intermetal dielectric (IMD) CMP. The ILD CMP process has been used to polish plasma-enhanced tetraethylorthosilicate (PETEOS) or high-density plasma chemical vapor deposition (HDPCVD) film on deposited silicon wafers. Figure 2.1 shows the ILD CMP process. As the manufacturing technique of the semiconductor device is developed, the number of levels in an interconnect technology is increased. To obtain the multilevel interconnection, the surface of wafer must be planarized to prevent topography roughness from growing with each level as shown in Table 2.1.

2.2 Rheological and Electrokinetic Behavior of Nano Fumed Silica Particle for ILD CMP

In the ILD CMP process, the most important factor is the characteristics of nano fumed silica slurry. The chemical interactions and physical properties of nano ceramic particulates must be considered to planarize the surface of wafer successfully. The dispersion stability of nano fumed silica slurry is directly related to the polishing rate (removal rate), the surface scratch, and the uniformity (within-wafer non-uniformity) of wafer surface across the whole wafer. Controlling the dispersion stability of nano fumed silica slurry is a key parameter in the ILD CMP process.

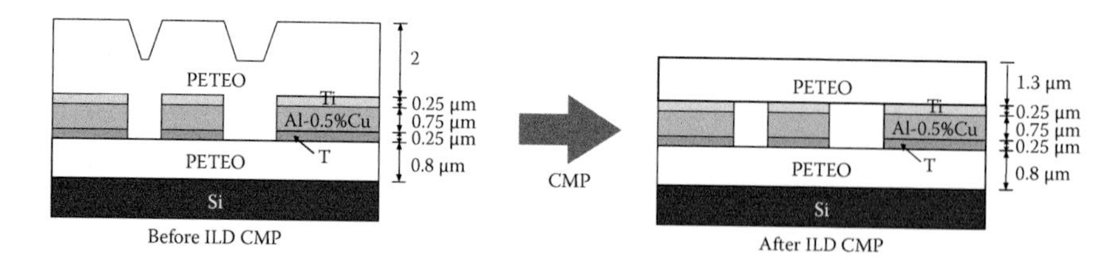

FIGURE 2.1 (See color insert) Schematic of ILD CMP process.

2.2.1 The Unique Behavior of Concentrated Nano Fumed Silica Hydrosols

Nanosize inorganic particles (i.e., below 100 nm) are gradually being incorporated into a broad range of advanced devices and applications. Some examples include silicon (Si) wafer polishing, planarization for semiconductor manufacturing (CMP process), electronic packages, ultrathin-film optical devices, advanced fuel cell catalysts, molecular conductors, and biochips. Recent evidence has indicated that classical colloid principles might not fully explain the complex behavior of concentrated nanosols.

According to the Derjaguin–Landau–Verwey–Overbeek (DLVO) theory, a cornerstone of modern colloid science, two types of forces exist between colloidal particles suspended in a dielectric medium: (1) electrostatic forces, which result from unscreened surface charge on the particle; and (2) London–van der Waals attractive forces, which are universal in nature. The colloidal stability and rheology of oxide suspensions, in the absence of steric additives, can be largely understood by combining these two forces (assumption of additivity).

There are several reports of the unique stability of nanosize silica hydrosols near the isoelectric point (IEP). The Canberra group experimentally discovered the existence of short-range forces that play an important role in the interaction process and must be added to those forces already accounted for by the original DLVO theory. These short-range interactions

TABLE 2.1
Roadmap for ILD CMP

Year of Production	2006	2007	2008	2009	2010	2011
DRAM ½ pitch (nm)	70	65	57	50	45	40
Flash ½ pitch (nm)	64	57	51	45	40	36
CMP Performance						
Dishing (A)	<500	<400	<300			
Erosion (A)	<1000	<500	<300			
Uniformity	5%	3%	2%			
Defect (μm)	<0.13	<0.10	<0.08			

are referred to as structural forces. Structural forces might explain some particular aspects of the stability behavior of silica nanosols, but they are insufficient to account for the apparent cooperative effects of solids loading and electrostatic found in the present study. Contrary to suspensions based on colloidal-size (100–1000 nm) silica and other inorganic oxides as reported in the literature, we found that the rheological behavior of concentrated electrostatically stabilized silica nanosols is counterintuitive with regards to the predictions based on a standard interpretation of DLVO theory. Despite the high surface charge density electrokinetic potential at pH 8, nano fumed silica particles not only showed unstable rheological behavior that would normally indicate an unstable or aggregated suspension (i.e., pseudoplastic high viscosity), but the rheology did not have the expected dependence on ionic strength. In this chapter, experimental measurements, DLVO calculations, and simple geometric considerations are used to understand the influence of solids loading and the electrical double layer on the rheological behavior of concentrated silica (20 nm) nanosols, and to compare their behavior with that of much larger silica microspheres, as well as like-sized nano-alumina, under similar conditions.

2.2.2 Electrokinetic Behavior of Nano Silica Hydrosols

By changing the pH, one can alter the magnitude (and sign) of the zeta (ζ) potential, whereas the addition of an inert electrolyte will affect both the magnitude of ζ and the electrical double-layer thickness. Thus, both pH and electrolyte concentration will directly impact colloidal stability in an electrostatically stabilized system. Figure 2.2 compares ζ potential and viscosity (at a shear rate of 26.4 s^{-1}) as a function of pH for the nanosized fumed silica and the silica microspheres suspensions. The average primary particle sizes were 20 nm for the nanosized fumed silica and 500 nm for silica microspheres. Even at a solids concentration of 20%, the silica microspheres exhibit a fairly constant and low viscosity across the entire pH range, whereas nanosized fumed silica exhibits a strong pH dependence at a volume fraction of 13.2% with an increase in viscosity near pH 7 in excess of 300mPa s. Figure 2.3 shows the effect of inert electrolyte concentration on viscosity as a function of shear rate for highly charged 13.2% nanosized fumed silica at pH 8.

Figure 2.2 indicates that for the silica microspheres, the ζ potential and viscosity both follow the expected behavior predicted by the classical DLVO theory. On the other hand, the nanosized fumed silica exhibits a discrepancy between the expectation of DLVO theory and the experimental results, that is, as ζ of the nanosized fumed silica increases, viscosity sharply increases. Hence, factors such as particle crowding, particle ordering, and electroviscous effects will also impact viscosity, in addition to aggregate or network formation.

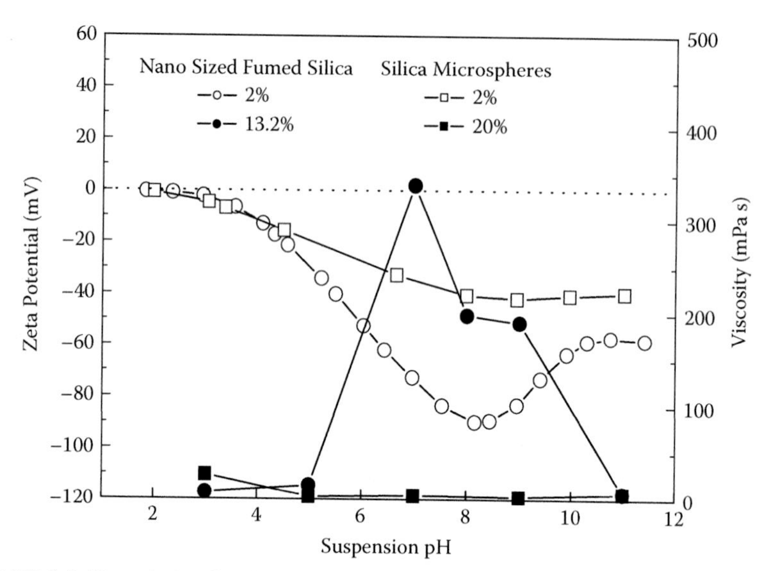

FIGURE 2.2 The relationship between zeta potential (open shapes) and viscosity (filled shapes) for silica suspensions as a function of suspension pH: nanosize A90 versus Geltech microspheres (G). Viscosity was determined at a shear rate of 26.4 s^{-1}. Particle volume fraction given in percent.

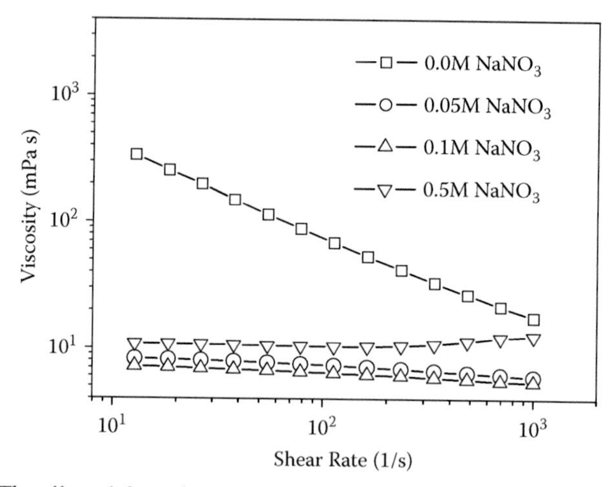

FIGURE 2.3 The effect of electrolyte concentration on the viscosity of 13.2% A90 silica at pH as a function of shear rate.

2.2.3 Geometric Consideration

To more properly analyze the results of Figures 2.2 and 2.3, it helps to first layout the physical dimensions of the system as depicted in Figure 2.4. The mean interparticle center-to-center separation distance (d_{c2c}) is defined as $d_p/\Phi^{1/3}$, where d_p is the primary particle diameter and Φ is the particle

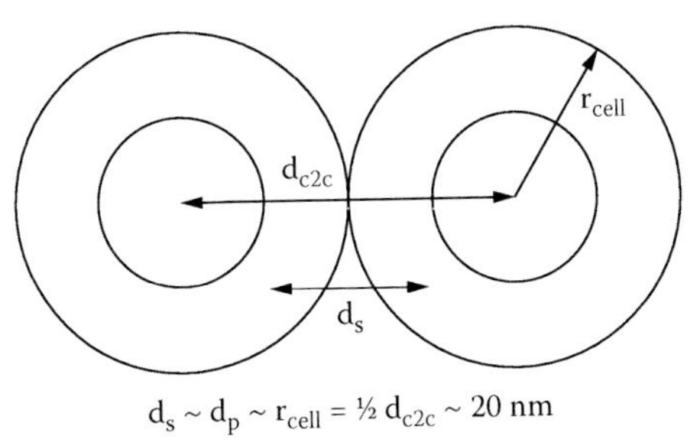

FIGURE 2.4 Diagram illustrating the relationship between average interparticle surface-to-surface separation distance, d_s, and other system dimensions, for a particle diameter d_p = 20 nm and Φ = 13.2%.

volume fraction. Then the mean interparticle surface-to-surface separation distance (d_s) is $d_{c2c}-d_p$. As Φ increases, the system dimensions, d_s and d_p, eventually become of comparable length ($d_s/d_p \sim 1$), which can lead to constrained motion and excluded volume effects. That is, other particles may be excluded from the interparticle space once the average separation distance is of the order of the particle size, thereby reducing the number of possible positions each particle is able to sample during Brownian motion. Furthermore, each particle with a surrounding volume of liquid defines a spherical cell. Figure 2.5 shows the average cell radius, $r_{cell} = d_{c2c}/2$, and d_s as a function of Φ and d_p. As d_p decreases or Φ increases, d_s becomes smaller. This has important implications for nanosize particles, and helps

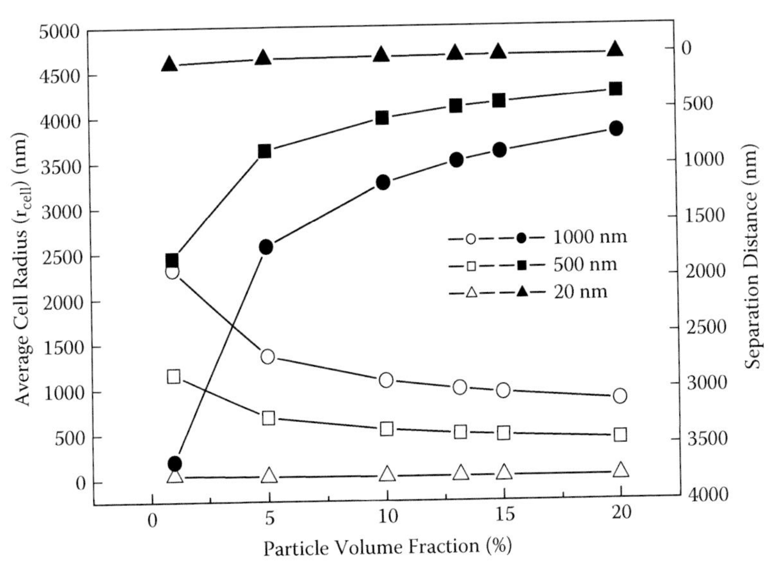

FIGURE 2.5 Calculated average cell radius (opened symbols) and surface-to-surface separation distance (filled symbols) as a function of particle volume fraction and particle size for silica.

explain why it is so difficult to obtain low-viscosity concentrated nanosols in aqueous systems. This explanation may not be immediately obvious since the critical Φ corresponding to $d_s/d_p = 1$ occurs at about 13%, irrespective of particle size.

However, the distance over which hydrodynamic and electrostatic forces act in solution is more or less independent of particle size at first approximation. As a result, when the average separation distance between particles is rather large, these forces dissipate before they can influence neighboring particles. As a result, particle motion is independent and the rheological behavior is Newtonian so long as the particles remain stable and do not aggregate. On the other hand, as the average separation distance is reduced, these forces begin to influence nearest neighbors, and the motion of nearby particles becomes coupled. Coupling leads to an increase in suspension structure, which provides an additional mechanism for viscous dissipation. In aqueous nanosols, the effects of electrostatic forces on structure can be particularly strong as d_p and d_s approach the length scale over which short-range repulsive interactions are active.

2.3 Particle Engineering for Improvement of CMP Performance

ILD CMP typically uses a fumed silica slurry dispersed in an aqueous medium at pH near 11. Fumed silica is a widely adapted abrasive for ILD CMP because of its inexpensive price, high purity, and colloidal stability. However, fumed silica is difficult to disperse in an aqueous system, and to control powder processing, because of a large specific surface area of 90 ± 15 m^2/g, making it very reactive. ILD CMP slurry was prepared at pH 11 to accelerate the chemical attack on the deposited PETEOS film on the wafer surface. But silica particles dispersed in aqueous media are partially dissolved at pH 11. Consequently, the removal rate decreased and microscratches were generated on the wafer surface as due to agglomeration of silica particles as surface potentials decreased.

2.3.1 Surface Modification of Silica Particle

As mentioned earlier, the dispersion stability of the slurry is directly related to CMP performance as removal rate—within-wafer non-uniformity (WIWNU), which is defined as the standard deviation divided by the average of remaining thickness after CMP, microscratch, and the remaining particle on the wafer. The agglomeration of particles causes low removal rate and the remaining particles of the deposited film surfaces. To

TABLE 2.2
Dissolution Amount of Si Ions with and without Surface Modification

	With Modification	Without Modification
Amounts	2,050 ppm	40,000 ppm
	0.070 ± 0.001 mol/L	1.370 ± 0.002 mol/L

avoid poor CMP performance, the dispersion stability of the slurry must be controlled by prevention of Si ion dissolution from SiO_2 film surface.

The amount of Si ions dissolution is found to be dependent on surface modification, which was confirmed by inductively coupled plasma–atomic emission spectrometer (ICP-AES) analysis. Table 2.2 shows the dissolution amount of Si ions with and without surface modification of fumed silica slurry. Without surface modification, the amount of Si dissolution was 1.370 ± 0.002 mol/L, whereas surfaces modified with poly(vinylpyrrolidone) (PVP) polymer yielded a dissolution of 0.070 ± 0.001 mol/L, almost 20 times less than the unmodified surface. Figure 2.6 represents the electrokinetic behavior of silica characterized by electrosonic amplitude (ESA) with and without surface modification. When PVP polymer modified the silica surface, dynamic mobility of silica particles showed a reduction from –9 to –7 mobility units (10^{-8} $m^2/V{\times}s$). Dynamic mobility of silica particles lacking this passivation layer shows that silica suspensions exhibit negative surface potentials at pH values above 3.5, and reach a maximum potential at pH 9.0. However, beyond pH 9.0, the electrokinetic potential decreases with an increasing suspension pH. This effect is attributed to a compression of the electrical double layer due to the dissolution of Si ions, which resulted in an increase of ionic silicate species in solution and the presence of alkali ionic species. When the silica surface was modified by

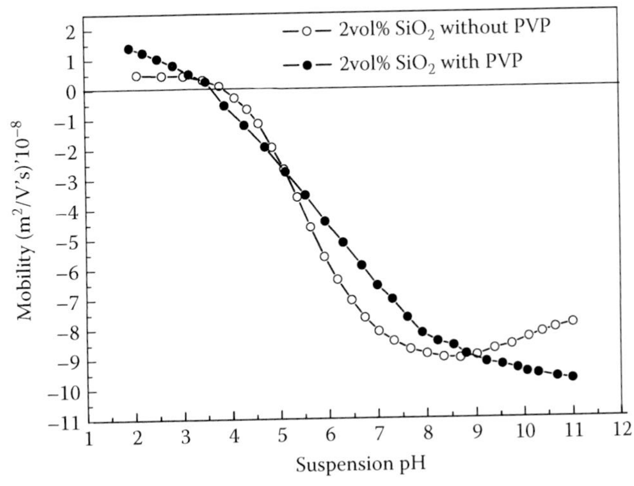

FIGURE 2.6 Electrokinetic behavior of silica suspensions with and without additive.

TABLE 2.3

Removal Rate and WIWNU with and without Modification

Wafer Number	Removal Rate (Å/min)		WIWNU (%)	
	With Modification	Without Modification	With Modification	Without Modification
1	2822	2873	3.57	8.09
2	2705	2767	3.75	8.62
3	2791	2813	3.96	9.43

PVP polymer, the decrease of electrokinetic potential above pH 9.0 disappeared, resulting in an increase of the stability of particles dispersed in the alkaline pH region.

2.3.2 Improvement of ILD CMP with Modified Silica Slurry

Table 2.3 shows the removal rate and WIWNU of silica slurry with and without surface modification. In comparing the results with and without the modification, the removal rate is similar, but the final WIWNU of modified slurry is better than that of nonmodified slurry. Removal rate and final WIWNU results were closely correlated to the surface potential, rheological behavior, and large-particle size distribution. In effect, the surface modification strongly influenced the suspension stability and, hence, the properties of wafer uniformity. The microscratch and remaining particles on the silicon wafer with and without the surface modified slurry are shown in Figure 2.7.

Microscratches and remaining silica particles on a wafer for the modified slurry is much lower than those for nonmodified slurry. PVP, which modifies the silica particles and plays a preventive role in dissolving Si ions, is thought to improve the suspension stability. Due to the surface

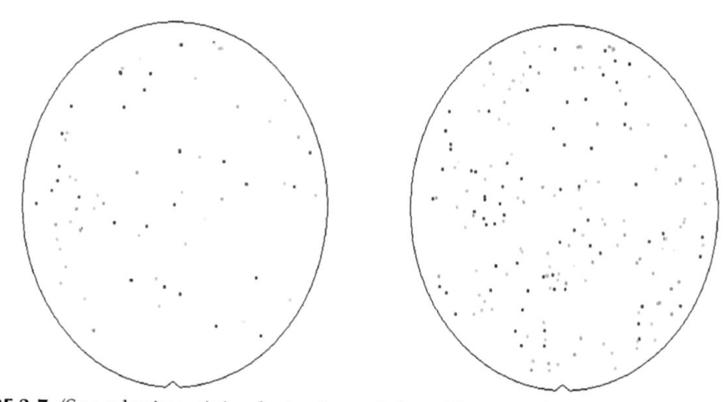

FIGURE 2.7 (See color insert) Analysis of remaining silica particles (particle size > 0.189μm) on silicon wafers after post CMP cleaning: (left) modified slurry, (right) nonmodified slurry.

modification, microscratches on the silicon wafer were decreased, as the improved suspension stability prevented the undesirable agglomeration. Additionally, as the reactivity of Si ion with the silicon wafer is much higher than that of silica particle, the stuck particles on the wafer surface decreased owing to reduction of the amount of Si ion dissolution in the case of the slurry modified with PVP.

2.4 PAD Dependency in ILD CMP

Lee et al. (2000) have systematically reported nanotopography impacts on oxide CMP, however, they quantified the impact using standard deviation of film thickness variation, which itself does not include lateral information. To develop an essential understanding of the relationship between the nanotopography and the film thickness variation after CMP, a treatment is needed not only for the amplitude but also for the wavelength component. Fukuda et al. (2000) qualitatively showed the pad dependency with a comparison between the line profiles of the nanotopography of wafers and the film thickness variations after oxide CMP. The pad dependency using two types of pad and the removal depth dependency is investigated with reiterated polishing followed by film thickness measurement. The result is examined by means of the spectral analysis method. As the comparison between soft pad and hard pad, the standard deviations of filtered film thickness variation of group-G are plotted in Figure 2.8. The standard deviation for hard pad test is two times as large as the one for soft pad test in the comparison for same removal depth. The reason for this trend can be understood from the difference of planarization length between two kinds of pad. That is, by the harder pad, more

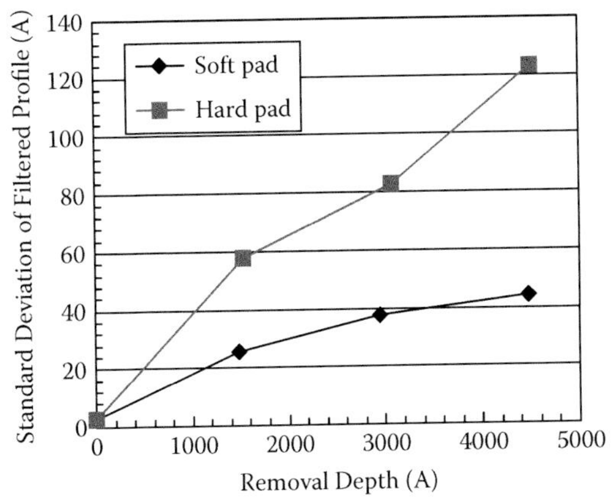

FIGURE 2.8 Standard deviations of the filtered film thickness variation before and after CMP. The pad dependency and removal depth dependency is plotted.

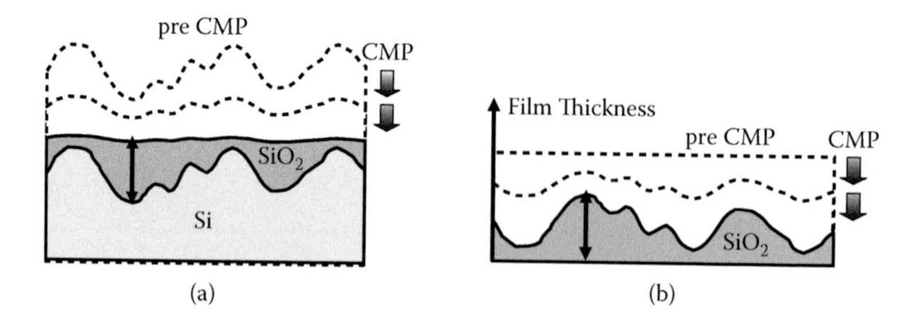

FIGURE 2.9 (a) A cross-sectional scheme for progressing polishing of oxide film. The surface of oxide film is gradually smoothed. (b) The transition of film thickness variation during polishing.

longer-wavelength components of surface waviness can be smoothed efficiently, so the planarization of the surface occurs more quickly with a smaller removal depth) than by the softer pad. Then the impact of nanotopography is more severe if compared for the same removal depth.

In Figure 2.8 the removal depth dependency is also shown. The increasing of the filtered film thickness variation with larger removal depth can be understood from Figure 2.9, where the oxide surface is gradually planarized during polishing.

Correlation between the standard deviation of nanotopography profile and the film thickness variation before/after CMP are shown in Figure 2.10 (for soft pad) and Figure 2.11 (for hard pad). It is reasonable that film thickness variations before CMP were independent of nanotopography. However, after CMP, the film thickness variation and nanotopography have positive correlation. Here two facts are pointed out.

1. Pad dependency: The slope in Figure 2.11 is steeper than in Figure 2.10 if compared for same removal depth.
2. Removal depth dependency: The slope gets steeper as the removal depth increases.

In other words, the nanotopography impacts on film thickness variation get more severe with the larger removal depth. These facts are consistent with the result shown in Figure 2.8 and it can be estimated how much the nanotopography depends on the film thickness variation.

The power spectral densities (PSDs) of nanotopography and film thickness variation are drawn in Figure 2.12. The PSDs of film thickness variation before CMP were much smaller than that on nanotopography with a factor of 100 or 1000 for all groups of wafer type. The PSD of film thickness variation is getting closer to that of nanotopography as the polishing goes on. This is consistent with the model in Figure 2.9. This trend is more remarkable for shorter wavelength region and it corresponds to descending transfer function mentioned in Section 5.3.2. As for the pad

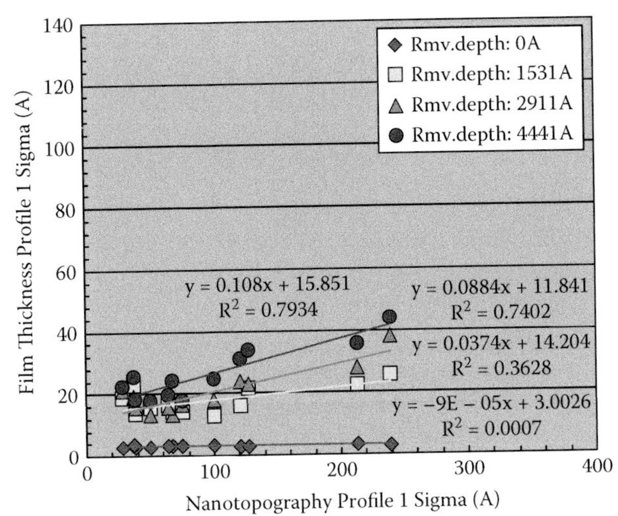

FIGURE 2.10 Correlation between standard deviations of nanotopography and film thickness variation for soft pad test.

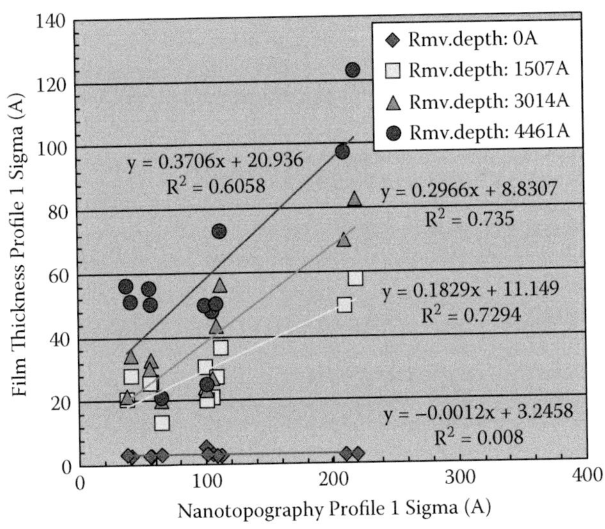

FIGURE 2.11 Correlation between standard deviations of nanotopography and film thickness variation for hard pad test.

dependency, with comparison for same removal depth, the PSD for hard pad is larger than that for soft pad in long wavelength. For some cases of large removal depth the PSD of film thickness variation exceeds that of nanotopography in short wavelength. It can be attributed to non-uniformity, which comes from other than nanotopography (e.g., pad non-uniformity), however it is not dominant.

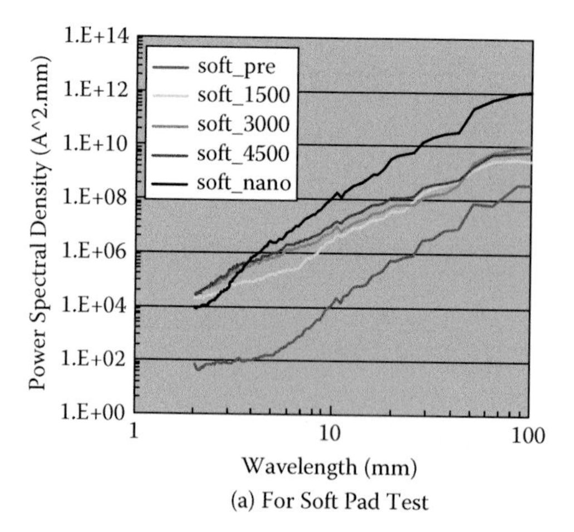

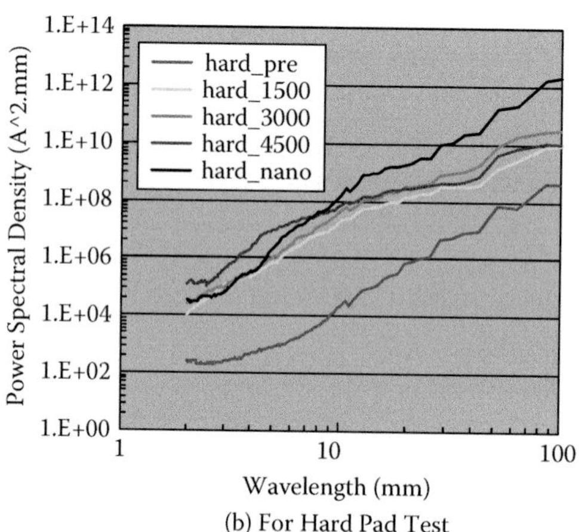

FIGURE 2.12 Power spectral densities for (a) soft pad test and (b) hard pad test.

2.5 ILD Pattern Dependencies

2.5.1 CMP Tool Dependency

CMP remains hampered by systematic and random interlayered dielectric (ILD) thickness variation at the wafer and die level. Pattern dependencies within the die, in particular, have been of concern for both manufacturability and product design.

Since the wafer- and die-level sources are deeply confounded, it is difficult to characterize the tool dependencies until these sources are decomposed. After application of variation decomposition techniques, the die-level variation can be analyzed for its pattern dependencies. Divecha et al. (1996) have shown that for similar polishing pad and processing conditions (e.g., platen speed, back/head pressure, and spindle speed) between the two tools, the die-level variation is similar and is

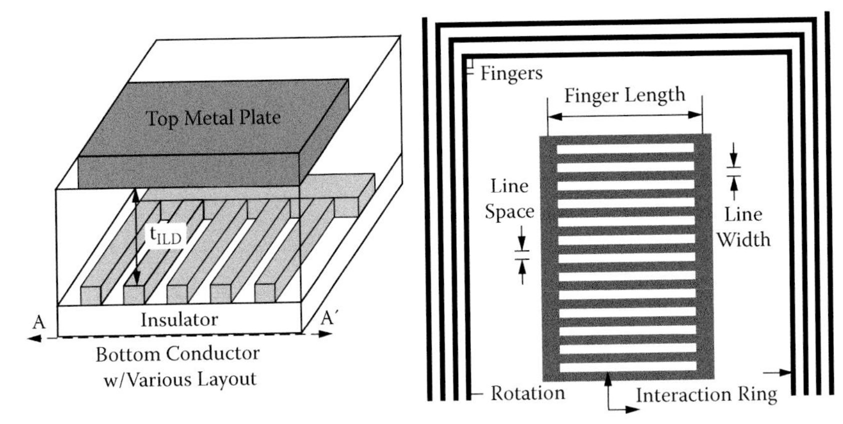

FIGURE 2.13 Capacitance test structures.

fundamentally dependent upon the underlying topographies, whereas a substantial wafer-level variation results.

Statistical metrology is a methodology for the systematic assessment and quantification of the sources of variation in a given semiconductor manufacturing process. The methodology requires a large number of measurements for statistical modeling. It also emphasizes the design of experiments to develop electrical test structures, use of short flow processes to ensure minimum variation in the final parameter from the confounding interactions between processing steps, and close coupling to technology computer-aided design (TCAD) tools necessary for extracting the desired parameters from electrical measurements. The test structure used in this experiment is a metal-to-metal capacitor to infer the ILD thickness as shown in Figure 2.13. The capacitor test structure has a uniform top electrode and bottom electrode consisting of various combinations of layout factors such as line width and spacing, finger length, the number of fingers, geometric orientation, and presence or absence of an interaction ring around the structure.

Combinations of six layout factors form a half-fractional factorial experiment yielding 32 unique structures. Four structures are put together in a subdie layout shown in Figure 2.14 with corresponding resistive structures to account for local line width variation. The subdie layout is replicated four times within the die to obtain spatial mapping. Figure 2.15 shows the 1.45 cm × 1.45 cm short-loop test die. The fourth quarter of the die includes large uniform density intensive structures to study the area dependence, and also serve as dense patterns in the die. Test wafers were processed in a short-flow Metal1–Metal2 process, with half of the wafers being polished on commercial tool A and the other half on commercial tool B. Slightly different slurries and pad conditioning techniques were used, but the polishing pads, back/head pressure, platen speed, and spin-

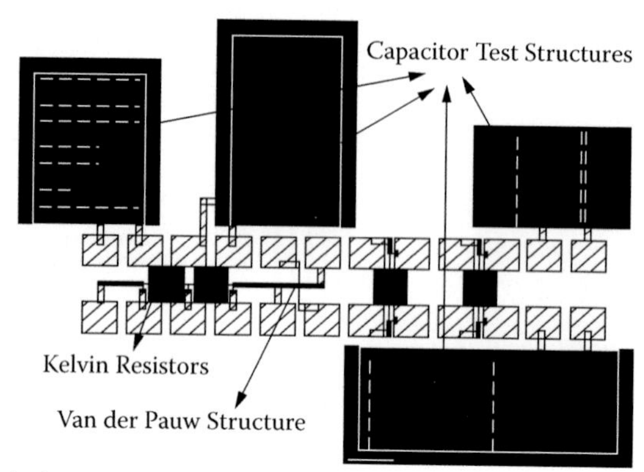

FIGURE 2.14 Probe layout.

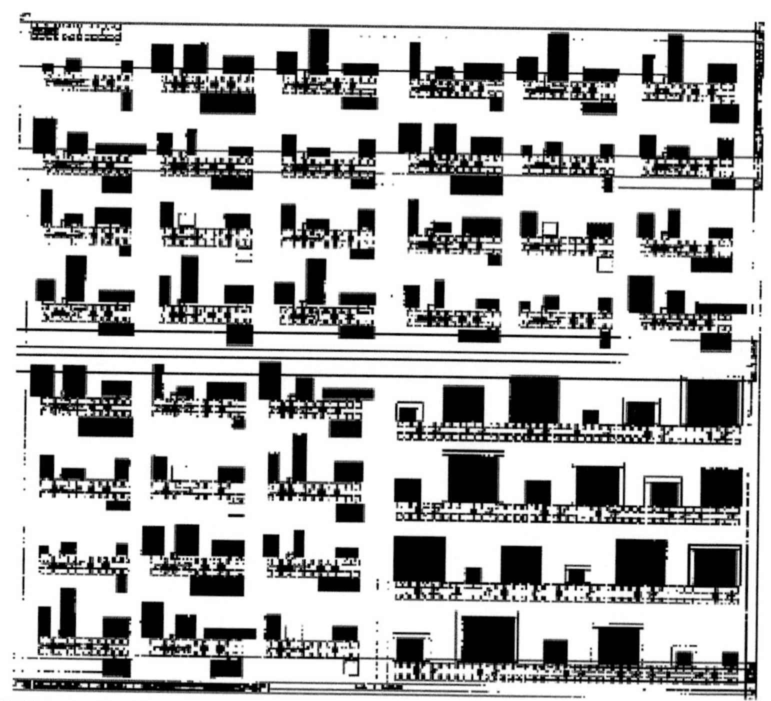

FIGURE 2.15 CMP/ILD thickness die layout.

dle speed were similar on the tools. The ILD thickness data were extracted from AC high frequency (100 kHz) capacitance measurements.

The ILD thickness variation sources can be categorized into wafer-level, die-level, die and wafer interaction, and residual terms. The wafer-level variation is often caused by process perturbation and drifts in equipment and consumables, and is relatively invariant of pattern density and other layout effects. On the other hand, the die-level variation is attributed to

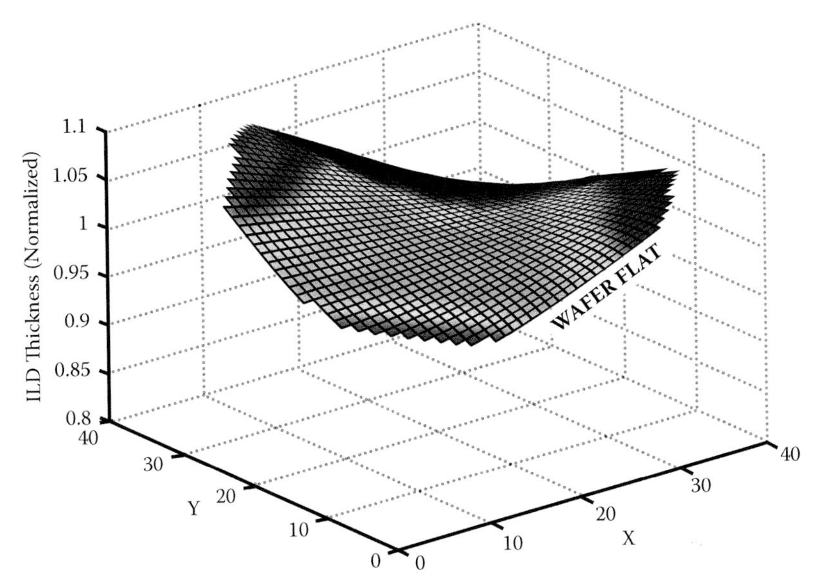

FIGURE 2.16 (See color insert) Wafer level variation for tool A.

the layout patterns within a die. Statistical methods were used to separate these two components. Figure 2.16 and Figure 2.17 depict the wafer-level variation extracted from a typical wafer polished on tool A and tool B, respectively. The effect of wafer edge and flats can be discerned from these figures. Both tools exhibit substantial wafer-level (or interdie) variation. These results are quite striking given that all the wafers from one lot were fabricated using the same deposition process. Clearly, the equipment factors, different pad conditioning techniques, slurry distribution, and other macroscopic physical effects during polishing contribute to this wafer-level non-uniformity.

Figure 2.18 and Figure 2.19 show the die-level variation held common between all dies on the wafer. The die-level (or intradie) ILD thickness variation pattern is found to be nearly identical. Both tools exhibit a similar pattern "signature" at the die level with the primary difference being relative attenuation in the magnitude of the variation. We attribute most of this attenuation to a difference in the total thickness of oxide removed. The denser structures on the fourth quarter of the die have less variation and are polished slower compared to the rest of the features. These results indicate that feature/pattern scale variation appears to be largely determined by pad and layout characteristics and are only weakly impacted by process conditions. Figure 2.20 shows a quantitative comparison of extracted die-level variation components for tool A versus tool B. The correlation coefficient for tool A versus tool B is 97%, and the magnitude of the slope indicates the difference in attenuation of the variation. A

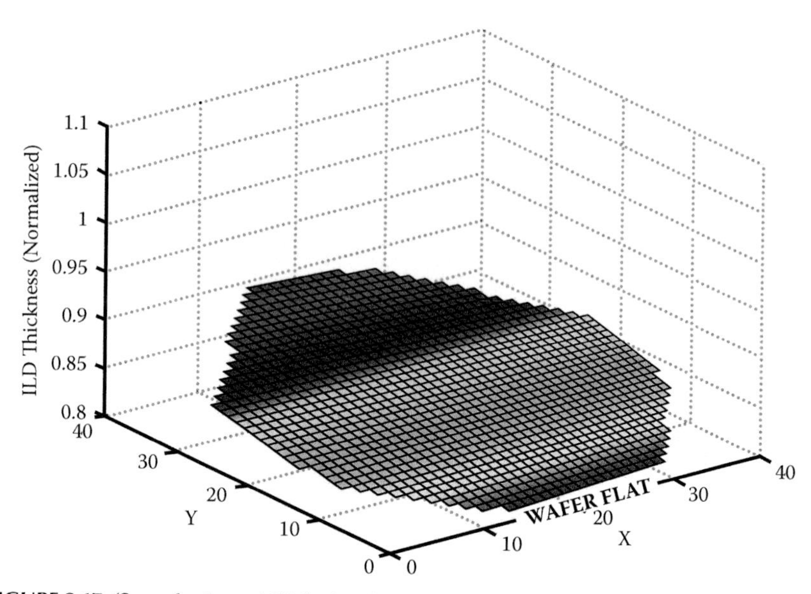

FIGURE 2.17 (See color insert) Wafer level variation for tool B.

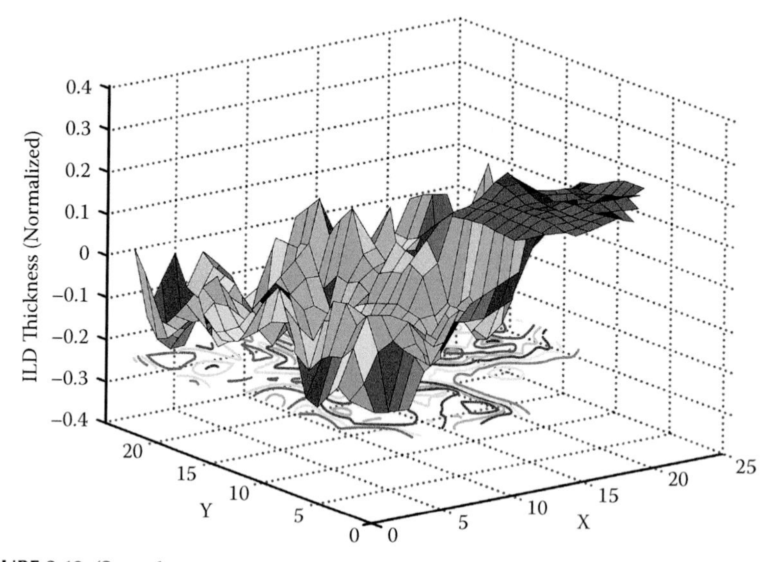

FIGURE 2.18 (See color insert) Die variation for tool A.

qq-norm plot also revealed that the residuals from the linear fit shown in Figure 2.20 are normally distributed. The explicit decomposition and modeling of wafer- and die-level variation, especially across different tools and consumable sets, can be extremely useful as part of a program to reduce pattern-sensitive effects in CMP.

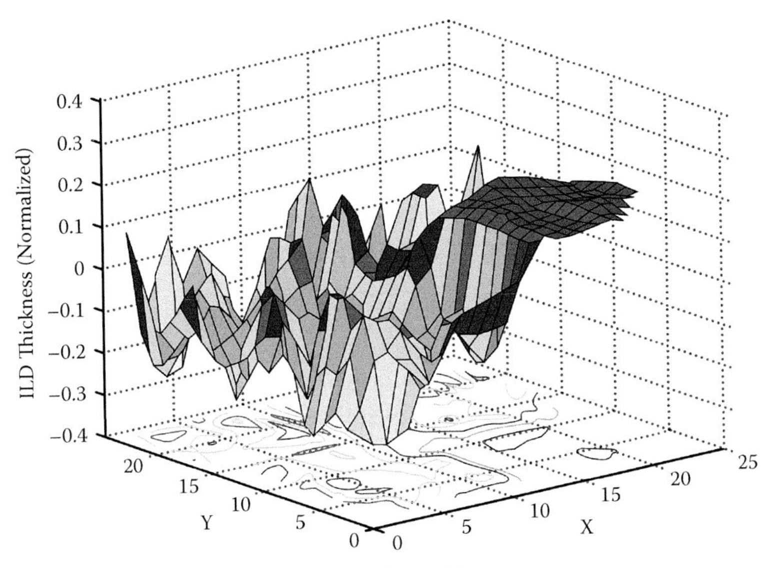

FIGURE 2.19 (See color insert) Die variation for tool B.

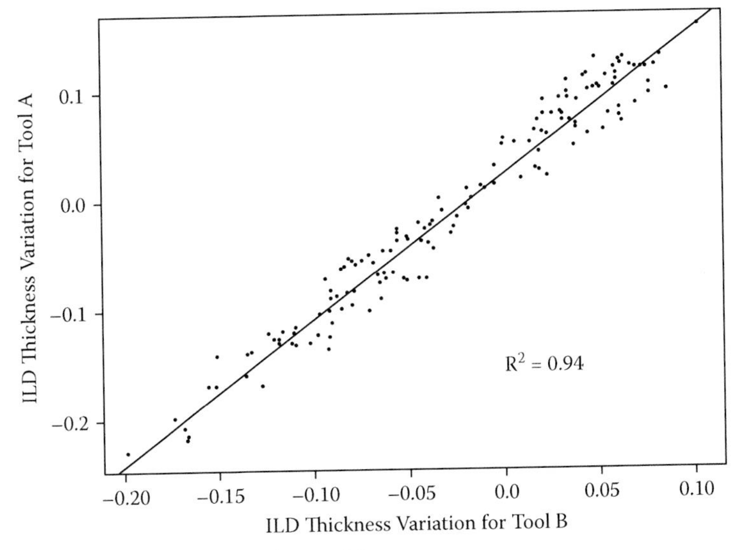

FIGURE 2.20 Die variation for tools A and B.

2.5.2 Pattern Density Dependency

Particularly, CMP characteristics are strongly dependent on the pattern density of the chip layout. Although CMP can planarize over long-length scales, the pattern density variation across a chip leads to large variation in global thickness across the die. The initial difference of layout pattern density between two regions creates a global step height across the two regions due to the difference in removal rates before the local patterns are planarized. The global thickness variation also impacts circuit performance: the long-range clock wires passing through the regions of different thickness result in different capacitances and may result in clock skew. Detailed understanding of the polishing process and subsequent improvements with better consumables and the process control, which minimize the pattern-density-dependent variation across a chip, must be employed. It has thus become a common practice to use dummy fill structures across a chip to minimize the pattern density variation. The dummy fill refers to the introduction of extra metal lines or blocks along with the actual metal interconnection. The main purpose of using the extra metal is to reduce the pattern-dependent ILD thickness variation across a chip by reducing the pattern density variation. Introduction of a dummy pattern may increase the circuit capacitance, thus it is important to minimize the using of the dummy pattern. In order to know how to include dummy patterns efficiently, preliminary polishing experiments should be done with specially designed wafers having various pattern densities and shapes.

Kang et al. (2001) investigated the characterizations of pattern-dependencies in ILD CMP using 8-inch SKW1 wafers designed by SKW Associates. The SKW1 density mask has structures with varying local pattern densities from 4% (lower-left corner) to 100% (upper-right corner) consisting of 25 blocks, as shown in Figure 2.21a. Each block is 4 mm × 4 mm with a fixed pitch of 250 μm, and the pattern density is increasing gradually at increments of 4%. Figure 2.21b shows the image of this layout of the die, and Figure 2.21c shows the cross-sectional view of the SKW1 dielectric CMP process characterization wafer.

The process conditions for this experiment are listed in Table 2.4. An IC1000/Suba IV K-Grooved stacked pad from Rodel Co. and an SS25 slurry from Cabot Co. were used. The thicknesses of prepolished and polished wafers were measured across the wafer by an Opti-Probe™ 2600DUV from Therma-Wave. To obtain within-die non-uniformity (WIDNU), 25 measurements per die were done over the metal (not between metal lines) near the center of each density block as shown in Figure 2.22. Five dies a wafer were measured to compare WIDNU at different positions on a wafer. To obtain within-wafer non-uniformity (WIWNU), 30 dies across the wafer measuring one site per die on the same density block of 52% were measured. In this case, not all 25 sites per die are necessary. Only one measurement per die is required on the same density block. The sampling scheme of these thickness measurements is depicted in Figure 2.23.

Figure 2.24 shows the removal rate variation as a function of pattern density for five dies with the positions shown in Figure 2.23a. This trend indicates that the removal rate decreases linearly as the pattern density increases. It can intuitively be explained by Preston's equation, $R = k_p p v$, where k_p is the Preston coefficient, R is the material removal rate, p is pressure, and v is relative velocity. As pattern density increases, the effective

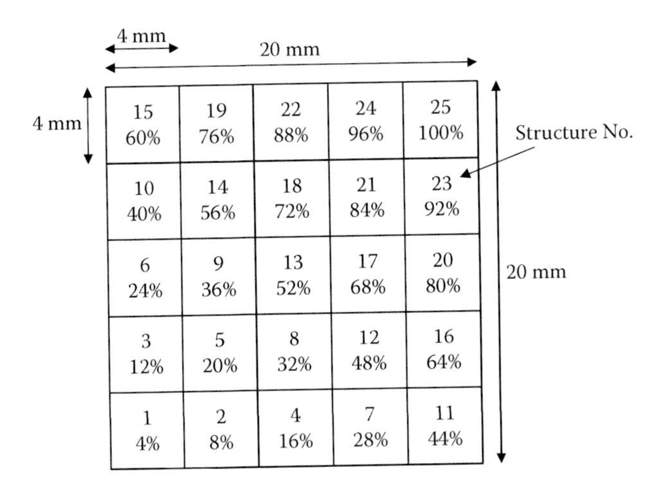

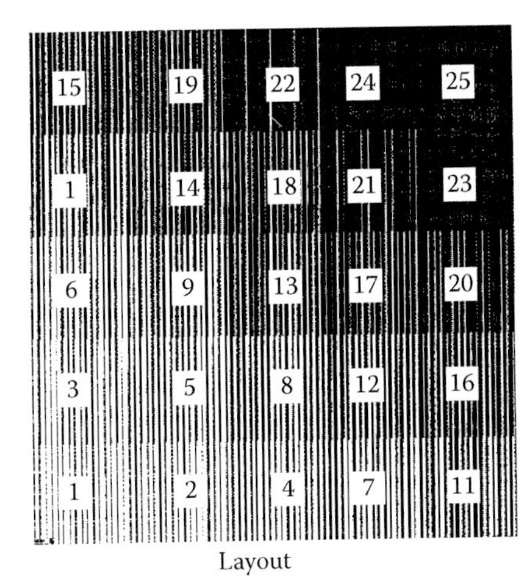

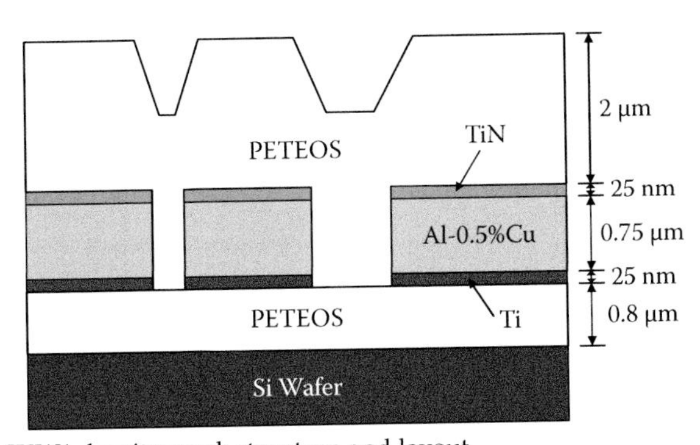

FIGURE 2.21 SKW1 density mask structure and layout.

TABLE 2.4
Process Experiment Conditions

Number	Down Force	Table Speed	Spindle Speed	Back Pressure	Slurry Flow Rate	Time
1	7 psi	30 rpm	50 rpm	0 psi	200 ml/min	246 sec
2	7 psi	30 rpm	50 rpm	0 psi	200 ml/min	295 sec

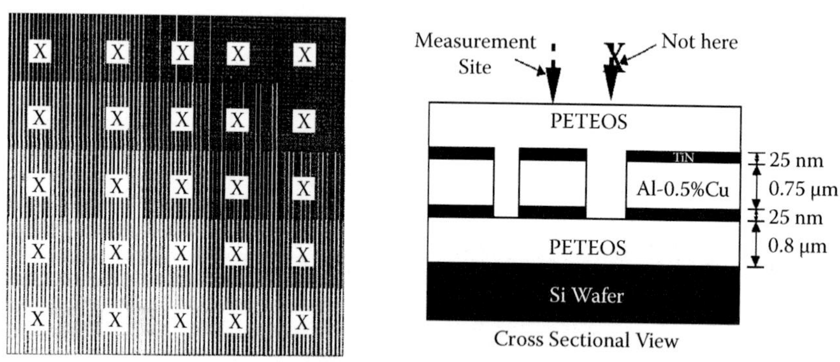

FIGURE 2.22 Measurement sites per die for SKW1.

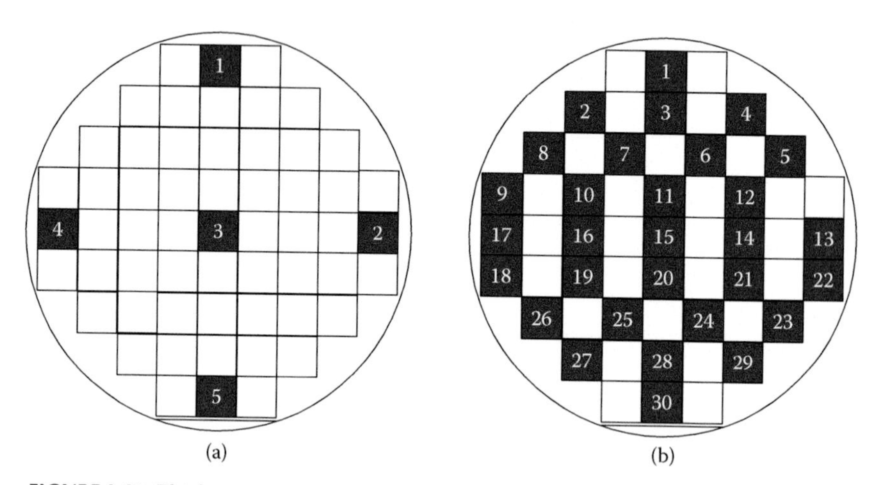

FIGURE 2.23 Thickness measurement sampling scheme: (a) five die measurements for die-level variation and (b) thirty die measurements for WIWNU.

contact area between the pad and wafer increases, and then the effective local pressure p becomes lower, resulting in a reduced removal rate. However, since process time in this experiment is longer than the time needed to completely remove step heights, the removal rates in Figure 2.24 are not directly given to real polishing rates for the pattern densities.

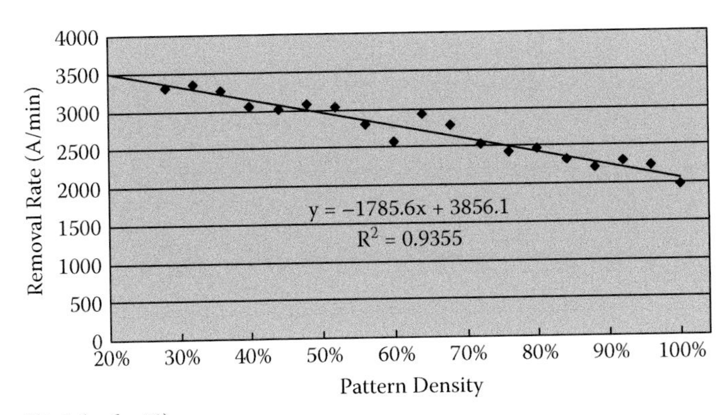

Die 1 (wafer #1)

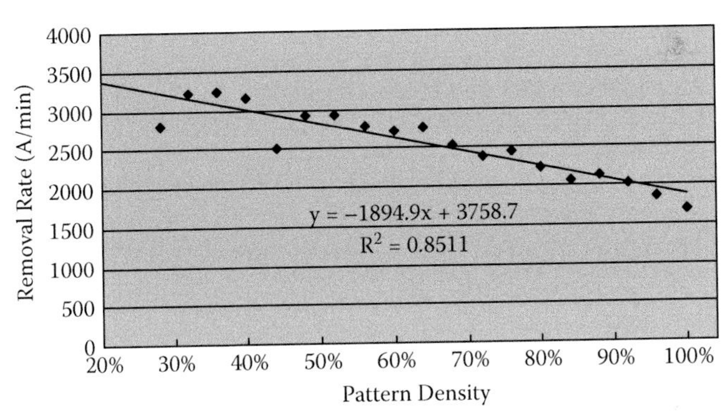

Die 2 (wafer #1)

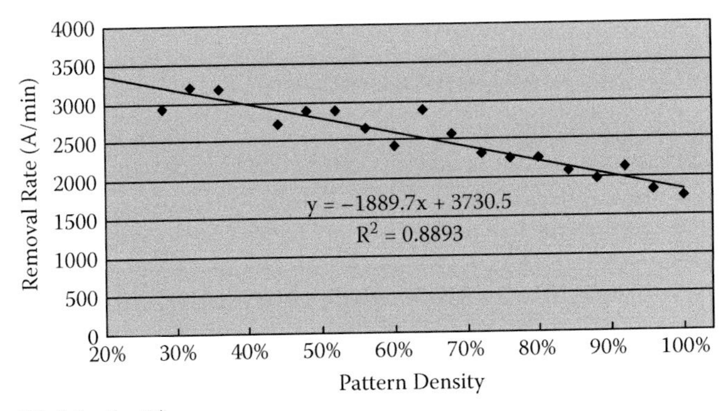

Die 3 (wafer #1)

FIGURE 2.24 Removal rate variation as a function of pattern density.

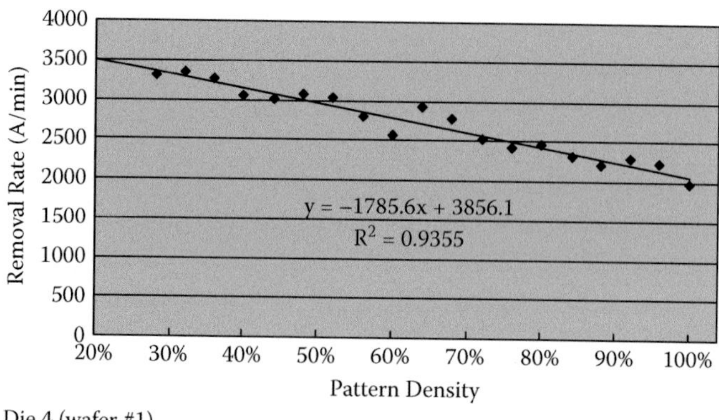

Die 4 (wafer #1)

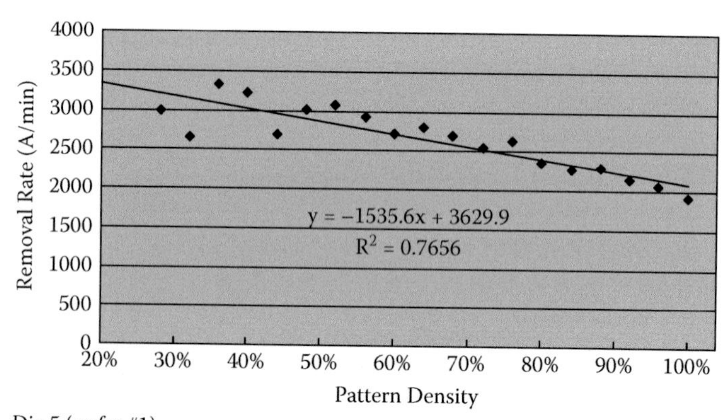

Die 5 (wafer #1)

FIGURE 2.24 (continued)

Figure 2.25a shows the removal rates for five dies located at different positions in a wafer. It represents that the removal rate of the center of the wafer is relatively lower than that of the edge. This result cannot be explained simply because this kind of non-uniformity can be caused by a lot of factors such as machine characteristics, machine conditions, consumables, and wafer shape. The WIDNU is about 20% as shown in Figure 2.25b, which is inevitable because there are various pattern densities in a die for the practice of pattern density effect. Therefore, we can see that the pattern density in a die must be kept uniform in real CMP process.

The average WIWNU for SKW1 wafers is 6.4% as shown in Figure 2.26. Intuitively, the more the removal thickness, the better planarity achieved; however, at the same time, the across-wafer final thickness non-uniformity becomes worse.

To predict the final oxide thickness in a pattern wafer, we used a model known as the MIT model proposed by Stine et al. (1997). Figure 2.27

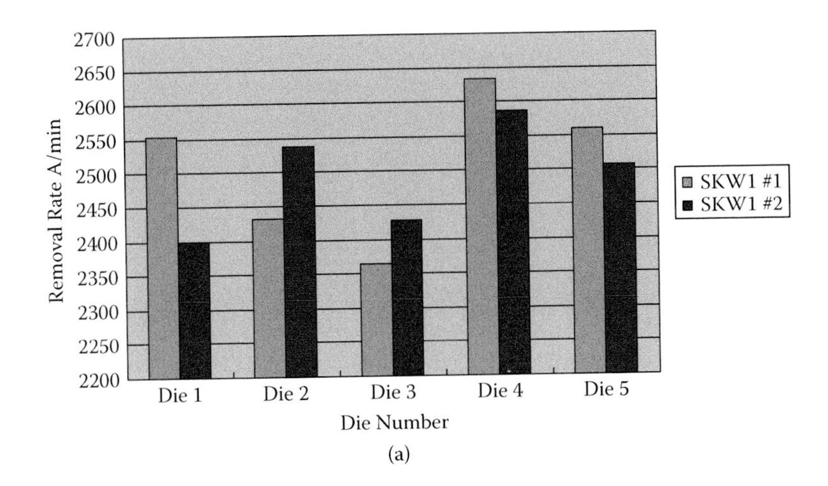

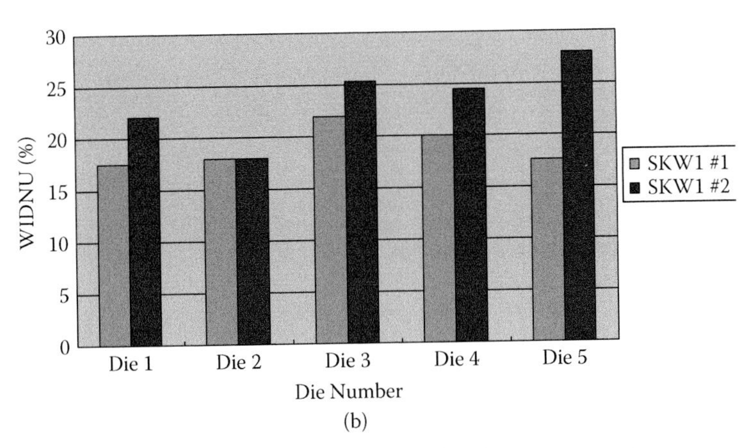

FIGURE 2.25 (a) Removal rates and (b) WIDNU for five die measurements.

defines the terms used in the model. In this model, the pattern density is considered as a function of time because the deposited film has a different surface area with the polishing time.

Preston's equation can be reformulated as:

$$\frac{dz}{dt} = -k_p p v = -\frac{K}{\rho(x, y, z)} \tag{2.1}$$

where K is the blanket-polishing rate and $\rho(x, y, z)$ is the effective pattern density. The equation is then solved for the oxide thickness z under the assumption that no "down area" polishing occurs until the local step, z_1, has been removed, after which the pattern factor is turned off. This is given by expressing the effective density as follows :

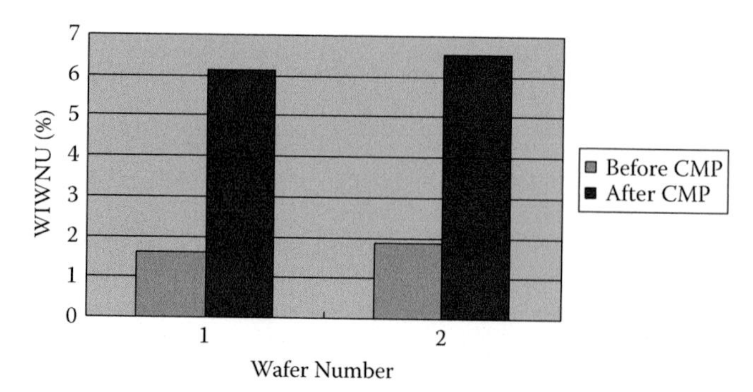

FIGURE 2.26 WIWNU for SKW1 wafers.

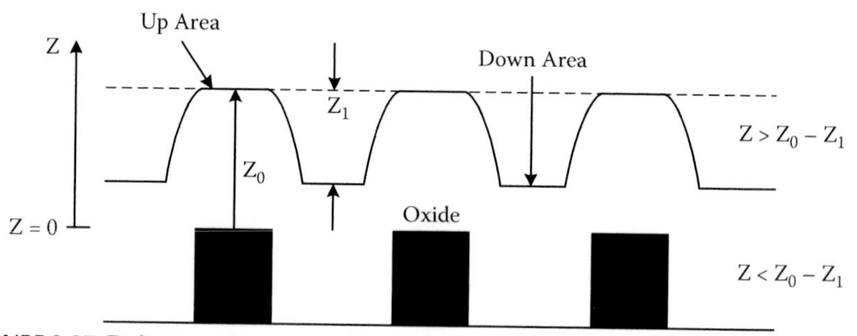

FIGURE 2.27 Definition of terms used in the basic model.

$$\rho(x,y,z)=\begin{cases}\rho_0(x,y,z) & z>z_0-z_1\\ 1 & z<z_0-z_1\end{cases} \tag{2.2}$$

In the determination of the effective density, $\rho(x, y, z)$, the effect of lateral deposition is accounted for by adding a bias term to the metal lines, which constitute the mask layout pattern. This ensures that the effective density is that of the final film profile and not the initial mask layout. It is assumed that the local pattern density is independent of the film thickness before the local planarity approximates the actual deposition profiles with a vertical profile. In reality, the effective density of the exposed surface depends on the height; it is possible to "time step" the profile evolution to account for such a time-varying density, but such detail is not essential for the prediction of final oxide thickness. The assumption makes it possible to express the final film thickness for any time, *t*, in a closed form as:

$$\rho(x,y,z)=\begin{cases}\rho_0(x,y,z) & z>z_0-z_1\\ 1 & z<z_0-z_1\end{cases} \tag{2.3}$$

Before the local planarity is achieved (i.e., while the local step height still exists), the final film thickness is inversely proportional to the effective local density. The film is assumed to be polished linearly at the blanket rate afterward. Based on this closed form of Equation 2.3, the residual oxide film thickness was calculated. Our simulation results agree well with the experimental data as shown in Figure 2.28. As a result, we can conclude that this model is very useful to correctly predict the remained thickness after CMP polishing.

The effects of the pattern density on CMP characteristics using 8-inch SKW1 wafers from SKW Associates, which were specially designed for the characterization of pattern dependencies in ILD CMP, were investigated. The removal rates for various pattern densities and uniformities were evaluated and analyzed after CMP. The experimental result shows that the removal rate decreases linearly as the pattern density increases and these different removal rates for pattern densities cause bad WIDNU. It shows that a dummy pattern must be employed to minimize pattern density variation. However, the introduction of a dummy pattern may increase circuit capacitance, thus it is important to minimize the addition of dummy patterns. Therefore, to limit the removal rates across a die within reasonable values, we must determine what range of the pattern density is available in the die at the target residual thickness. Using a simple model that can take pattern density into consideration, the remaining oxide thickness was calculated and compared with the experimental data.

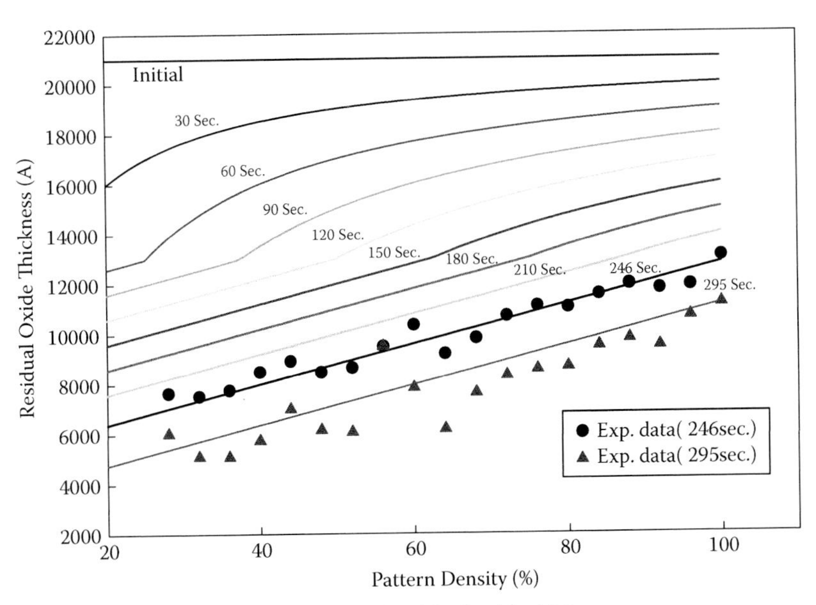

FIGURE 2.28 Experimental and calculated residual oxide thickness.

3

Shallow Trench Isolation CMP

3.1 Requirement for High Selectivity Slurry

Shallow trench isolation (STI) is a relatively new technique that is replacing local oxidation of silicon (LOCOS) for the manufacture of 64 MB semiconductor devices with a linewidth below 0.25 μm. Figure 3.1 shows the STI CMP process.

The STI process is defined as: (1) making a shallow trench to isolate active device regions physically, (2) depositing silicon nitride (Si_3N_4) on oxide films as a stopping layer, and (3) depositing oxide films on the trench. Generally, the STI process has the relative capability, compared to the LOCOS process, not only to deposit dielectrics to fill trenches isolating the active region at low temperature but also to prevent bird's beak and dimensional limitations. That is, STI CMP is essential for separation between transistors. If examining STI CMP's road map (Table 3.1) from this vantage point, the removal rate of oxide needs high selective slurry such that the nitride removal rate must increase and the number of scratches must decrease rapidly.

The key ingredients to a successful STI process are the achievement of well-dispersed abrasive ceramic particles having high oxide-to-nitride selectivity and producing few microscratches on the wafer. Silica slurry had been conventionally used in the STI CMP process, however, ceria (CeO_2) slurry with high oxide-to-nitride selectivity has been introduced as the thickness of silicon nitride film is decreased by design rule restrictions.

STI CMP processes with conventional oxide polishing slurries require either a reactive ion etching (RIE), etch back preplanarization step, or very tight control of the CMP process. Compared with other abrasive slurries, ceria slurry has a good selectivity between silicon oxide and silicon nitride. There are both chemical and mechanical interactions between ceria particles and wafer film during polishing. Nitride film is mainly affected by the chemical factors and the nature property of CeO_2 abrasive particle. Ceria slurries offer improved oxide-to-nitride selectivity for planarizing the trench fill material while utilizing the nitride film as the polishing stop layer. The oxide-to-nitride selectivity is a very important factor in

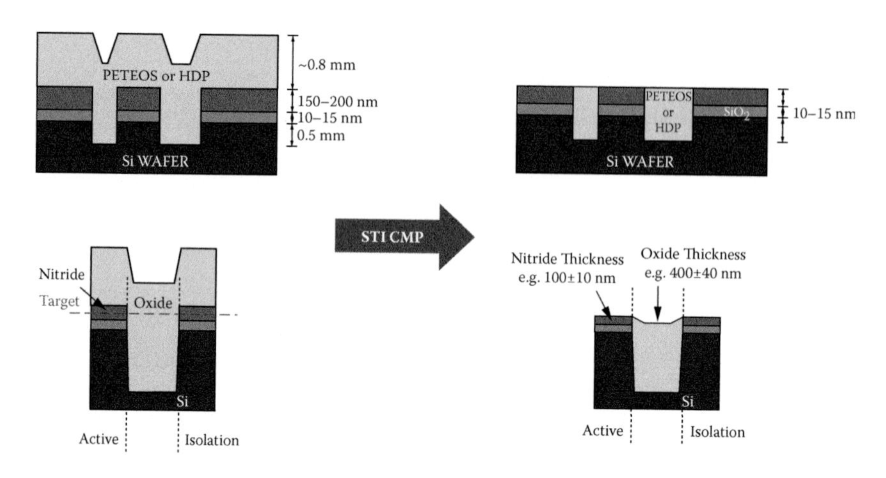

FIGURE 3.1 (See color insert) STI CMP process.

TABLE 3.1
Roadmap for STI CMP

Year of production	**2006**	**2007**	**2008**	**2009**	**2010**	**2011**
DRAM ½ pitch (nm)	70	65	57	50	45	40
MPU ½ pitch (nm)	78	68	59	52	45	40
CMP Performance						
Oxide removal rate [Å/min]			3000 ± 500	5000 ± 500	5000 ± 500	5000 ± 500
Nitride removal rate [Å/min]			50 ± 10	100 ~ 500 ± 5	100 ~ 500 ± 5	100 ~ 500 ± 5
Selectivity (oxide vs. nitride)			>35	>10 ~ 50	>10 ~ 50	>10 ~ 50
Remaining particle Size [nm]			≥90	≥65	≥65	≥45
Remaining particle [cm^{-2}]			≤0.35	≤0.17	≤0.17	≤0.17
Remaining particle [# wafer]			≤238	≤116	≤120	≤115
Scratch Count [#/wafer, 200 mm]				**≥1 μm**		
Oxide dishing [Å]			400	400	300	200
Nitride erosion [Å]			50	100	100	100
Metal contamination [ppma]			<1	<1	<0.5	<0.2

the STI CMP process. It can significantly affect the CMP induced defects, such as erosion or dishing, and also be important for endpoint detection. Figure 3.2 shows the effect of overpolishing on the characteristics of the device. Overpolishing due to nitride erosion or oxide dishing may cause degradation of device properties.

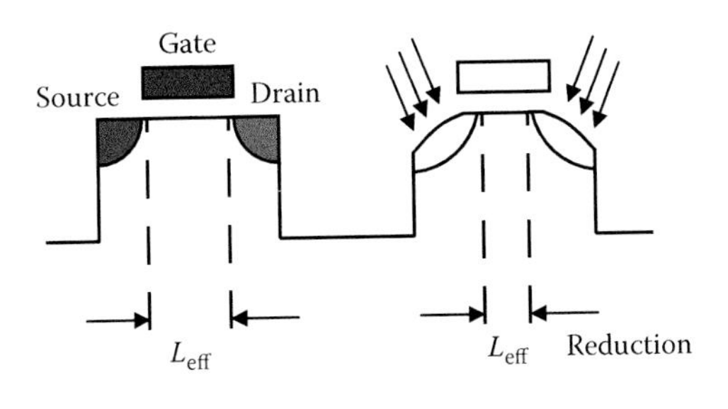

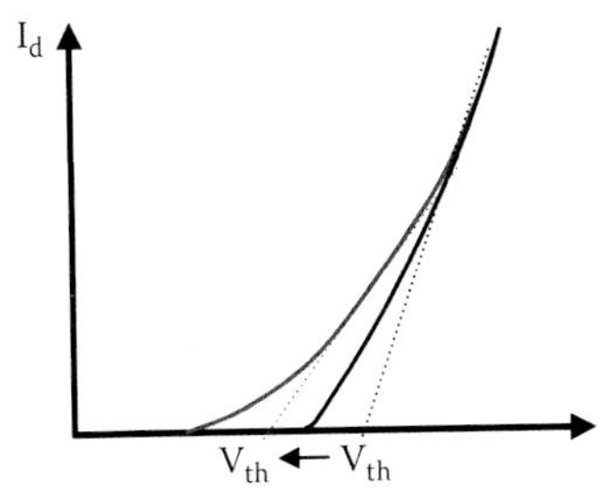

FIGURE 3.2 (See color insert) The effect of overpolishing on the characteristics of the device: (top) the decrease of gate length by overpolishing; (bottom) the shift of threshold voltage in the device.

If the effective gate length (L_{eff}) decreases, subthreshold drain current (I_D) may increase as shown by:

$$I_D = \mu_D \left(\frac{z}{L - y_s - y_D} \right) \frac{aC_i}{2\beta^2} \left(\frac{n_i}{N_A} \right) \left(1 - e^{-\beta v_D} \right) e^{\beta \psi_s} \left(\beta \psi_s \right)^{1/2} \tag{3.1}$$

and the threshold voltage (V_{th}) shifts as shown by:

$$\Delta V_T = \frac{qN_A W_m r_j}{C_i L} \left[\left(\sqrt{1 + \frac{2y_s}{r_j}} - 1 \right) + \left(\sqrt{1 + \frac{2y_D}{r_j}} - 1 \right) \right] \tag{3.2}$$

In general, two mechanisms can be applied to improve the selectivity between plasma-enhanced tetraethylorthosilicate (PETEOS) and Si_3N_4 during polishing the pattern wafer. One is chemical control using a surfactant to reduce the removal rate of Si_3N_4 and the other is mechanical control improving the physical properties of ceria particles to enhance the removal rate of PETEOS.

3.2 Particle Engineering of Ceria Nanoparticles and Their Influence on CMP Performance

The ceria particle is considered to be one of the best glass/SiO_2 polishing abrasives. This is suggested to be due to the reaction between ceria and SiO_2 film, which results in the formation of a chemical "tooth" between the silica surface and the ceria particles, and induced localized strain in the glass with particle movement. As a consequence, the Si–O–Ce bonds can be rapidly removed by the mechanical force generated by a pressed pad and abrasive particles. This physicochemical reaction leads to the high removal rate of a SiO_2 film by ceria particles. The physicochemical properties of ceria particles, such as crystallinity, particle roughness, and morphology, depend on the synthesis methods of cerium oxide. In this section, the influence of the ceria particles synthesis method on PETEOS and chemical vapor deposition (CVD) nitride films removal rates are presented.

3.2.1 Physical Properties of Ceria Particles

As the solid-state displacement reaction method and wet chemical precipitation method were employed for synthesizing ceria powders, the characteristics of ceria properties showed different features in several experiments. Figure 3.3 shows the morphology of ceria particles observed by a high-resolution scanning electron microscope (SEM; S900, Hitachi, Japan) and transmission electron microscope (TEM; JEM-2010, JEOL, Japan). In these figures, the ceria particles have a polyhedral shape. Both of the powders have nearly the same size. The primary particle size is ≈40 nm. However, the difference in crystal shape of the ceria particles was found in TEM analysis. Figure 3.4 shows x-ray diffraction (XRD) profiles of ceria powders produced by precipitation. The XRD data of the synthesized particles shows characteristics of CeO_2 with a typical fluorite structure. Since

FIGURE 3.3 SEM and TEM micrographs of CeO_2 particles: (a) SEM (100,000 magnification); (b) TEM (300,000 magnification).

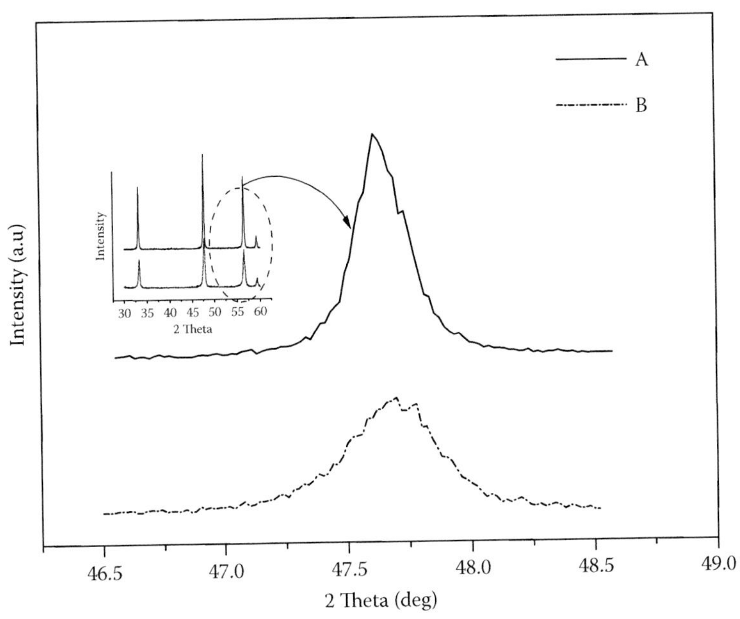

FIGURE 3.4 X-ray diffraction pattern of CeO_2 powders synthesized by precipitation method.

the starting cerium salt was $Ce(NO_3)_3$, it required oxidation of Ce^{3+} to Ce^{4+} in the solution. In this system, there is a possible cause for this oxidation. According to the Lewis definition of acids and bases, Ce^{3+} is a Lewis base and Ce^{4+} is a Lewis acid. Basic solution, therefore, favors Ce^{4+} compared to Ce^{3+}. The crystallite size was calculated from the Scherrer formula

$$D = 0.9\lambda \;/\; (\beta \cos\theta) \tag{3.3}$$

where λ is the wavelength of the x-rays, θ is the diffraction angle, and β is the half-width.

Average crystallite size of CeO_2 calculated by the Scherrer equation from the XRD line broadening was 46 nm for powder A and 34 nm for powder B. The crystallite size increases as the calcined temperature increases.

3.2.2 STI CMP Performance with Ceria Slurries

Figure 3.5a shows the result of CMP field evaluation. Average PETEOS removal rate of slurry A was 2883Å/min and B was 672Å/min. The within-wafer non-uniformity (WIWNU) shows that ceria slurry B (0.7%) is better than ceria slurry A (1.9%). Average nitride removal rate of slurry A was 51Å/min and B was 44Å/min as shown in Figure 3.5b. Thus, oxide-to-nitride selectivity was 56 for ceria slurry A and 15 for ceria slurry B. CMP field evaluation of ceria slurries having different crystallinity showed

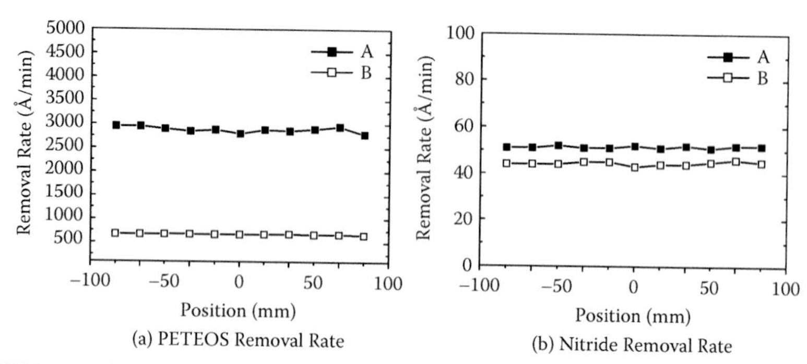

FIGURE 3.5 The result of CMP field evaluation: (a) profiles of PETEOS removal rate; (b) profiles of nitride removal rate.

that slurry A had better crystallinity and smaller pore size and exhibited a higher removal rate of PETEOS than B. Ceria slurry A showed higher removal rate and better planarization than slurry B. The oxide removal rate can be influenced by two CMP processing parameters: mechanical grinding and chemical interaction. These mechanisms play simultaneous roles in polishing. Concerning the chemical interaction between PETEOS and ceria slurry, Si–O–Ce bonding on surface is a dominant mechanism. During polishing of the PETEOS film, the SiO_2 surface would first react with CeO_2 particles and a multiple number of chemical bonding Si–O–Ce is formed on the surface.

Then mechanical tearing of Si–O–Si bonds leads to removal of SiO_2 or $Si(OH)_4$ as monomer lumps, then the lumps are released from the CeO_2 particles downstream. Highly crystallized ceria particles have a great tendency to form a bonding between Ce and Si, increasing the oxide removal rate. Ceria particles in B are unlikely to have less completed oxidation during the wet chemical precipitation and have less hard and less rigid surfaces. Therefore, B grains might have little effect in physical polishing with PETEOS film. Ceria particles in slurry A has almost fully crystallized on the surface after calcination at 800°C. CeO_2 surface of A would supply more potential site to react and bond between Ce and Si than that of B. Therefore, slurry A would interact with the oxide layer much more easily and readily and hence slurry A has a higher removal rate of PETEOS.

3.2.3 Influence of Crystalline Structure of Ceria Particles on the Remaining Particles

Figure 3.6 shows the morphology and primary particle size of the ceria particles after mechanical milling. The particle size distribution was found to be noticeably different between samples based on the TEM images. The portion of small-sized particles (<20 nm) in sample B and C were larger than sample A. The initial crystallite size of all as-calcined samples

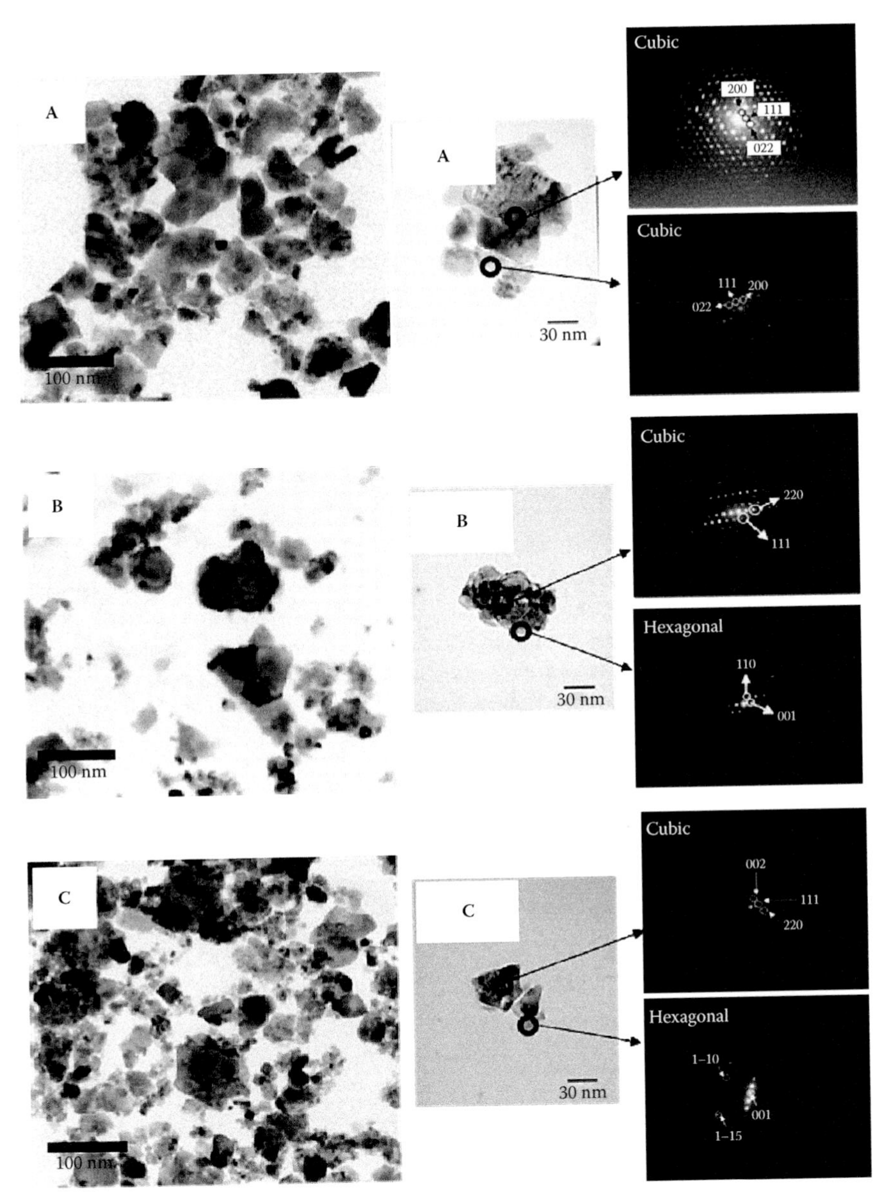

FIGURE 3.6 (See color insert) TEM images and nano beam diffraction.

calculated by the Debye–Scherrer equation was about 26 nm. However, the crystallite sizes of the particles after mechanical milling became 25, 21.5, and 20.3 nm for samples A, B, and C, respectively, which leads to a significantly different specific surface area for each particle group. After mechanical milling, the particles in all the samples had a similar size of 240–260 nm. Therefore, the particles of samples B and C, which were calcined at low oxygen concentration, can be considered to be composed of a large number of small crystallite (20.3 and 21.5 nm), while the particle of sample A, which was calcined at a higher oxygen concentration, is composed of relatively larger crystallite (25 nm).

Using the nano beam diffraction pattern of the high-resolution TEM, the crystalline structure of the ceria particles was investigated. The nano beam diffraction patterns of the particles (Figure 3.7) indicate that for sample A all particles had the cubic fluorite phase of cerium oxide, while for samples B and C, which were calcined at low oxygen concentration, hexagonally structured particles were included, especially in the smaller particles. The calcination process from cerium carbonate to cerium oxide consists of a five-step mechanism, including the mass transfer of the reacting agent (oxygen) from the bulk atmosphere to the periphery of carbonate, its diffusion through the pore channels of the carbonate, adsorption, reaction with the cerium carbonate, and the desorption of the reaction by-product (carbon dioxide). In this reaction mechanism, several factors influence the physical properties of the synthesized particles during the calcination process. According to previous reports, a low oxygen concentration results in a hexagonal phase cerium oxide rather than the cubic phase due to the insufficient oxidation of Ce^{3+} to Ce^{4+}.

In the absence of external oxygen supply, the hexagonal CeO_x phase is reported to be observed between 400°C and 500°C and the oxidation of CeO_x is completed above 800°C, which results in the transformation from hexagonal Ce_2O_3 to cubic CeO_2. Therefore, it can be considered that the hexagonal phase (Ce_2O_3) in samples B and C was formed due to the

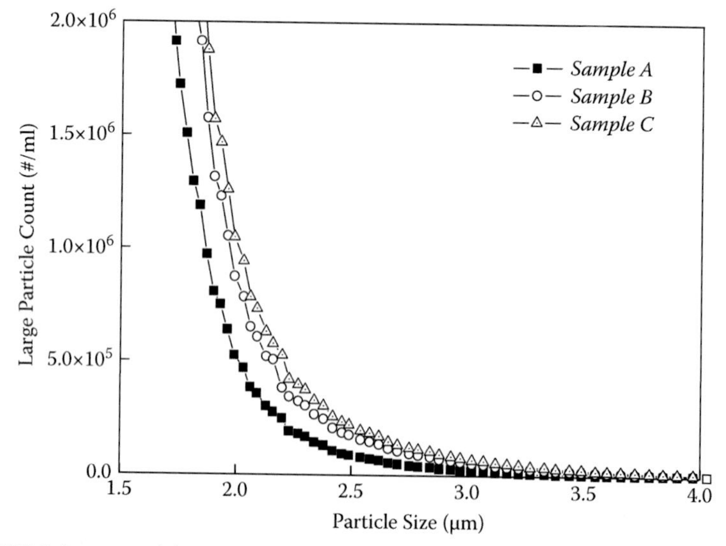

FIGURE 3.7 Large particle counts by Accusizer II.

insufficient oxidation of Ce^{3+} to Ce^{4+}, whereas the cubic phase (CeO_2) in the sample A was obtained by sufficient oxygen supply.

To investigate agglomeration in the slurries, we measured the number of large particles of over 1 μm per unit volume by using a slurry particle counter; the results are shown in Figure 3.8. Samples B and C with a hexagonal crystalline structure have more large agglomerated particles than sample A. Since the Ce^{3+} and oxygen vacancies on the surface of the hexagonal ceria particle are linked to anionic vacancies with hydroxyl groups, chemisorption and agglomeration easily occurred with the neighboring cation-species and bared ceria surface. Moreover, samples B and C have too many small particles, which were confirmed to be mainly hexagonal crystalline structure (Figure 3.7). The van der Waals attractive force is well known to be increased with decreasing particle size; therefore, it can be considered that large agglomerated particles in sample B and C are attributed to the presence of small-sized particles.

CMP was performed with the ceria slurry, which was prepared by adding an adequate amount of commercially available anionic acrylic polymers (PMAA). Cubic CeO_2 contained (sample A) and CeO_2 contained with included hexagonal Ce_2O_3 particles (sample C) were used as abrasive particles. The removal rate trends along the radius and WIWNU of the oxide film are shown in Figure 3.8. In Figure 3.9, the removal rate of oxide film shows a remarkable difference between samples A and C. The removal rate of the oxide film was dependent on the crystallinity of the ceria particle. The polishing of the oxide film was mainly affected by its chemical interaction between the ceria particles and the oxide (SiO_2) film, which were reacted with the hydrated surface to form covalent bonds such as

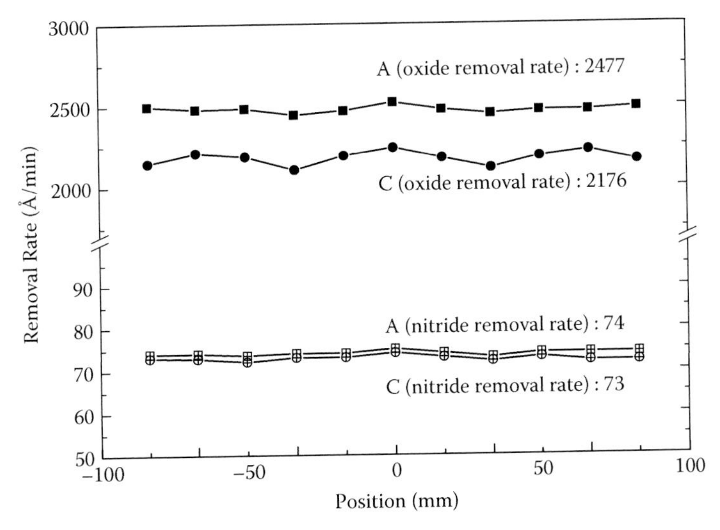

FIGURE 3.8 Removal rate trends.

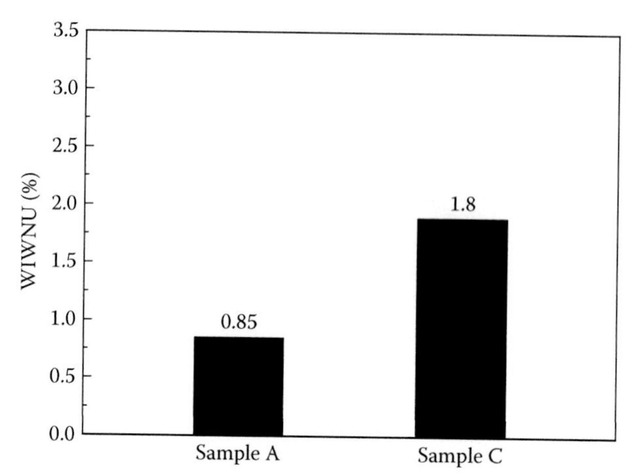

FIGURE 3.9 Within wafer non-uniformity.

Ce–O–Si, and then pulled off the oxide lumps. As shown in Figure 3.6, the particles of sample A have a larger crystallinity and narrower particle-size distribution than the particles of sample C. Since the hexagonal structured ceria particles and the agglomerated particles of sample C were easily broken apart during the CMP process, these particles do not penetrate the viscous layer on the oxide film. Thus, the removal rate of sample C was low, resulting in poor oxide-to-nitride selectivity. On the other hand, the surface of the silicon nitride film during polishing is passivated with an adsorptive surfactant in the slurry, which prevents the abrasive from directly contacting the film surface. Hence, the removal rates for the Si_3N_4 film were not influenced by the crystalline structure of the ceria particle. In addition, the slurry with agglomerated particles was hardly propagated over the whole wafer surface due to the poor stability. Thus, as shown in Figure 3.9, sample C has a higher WIWNU and a lower removal rate of wafer edge position than sample A.

During the CMP, the agglomerated particles were easily stuck to the wafer surface by the small interactive force between the abrasive and oxide film. It is these sticking particles in particular that induce the surface scratches on the wafer due to the compressive and shear forces between the wafer and pad. Therefore, agglomerated particles are a major cause for the residual particles and the microscratches in the CMP process. The maps of the residual particle counts and scratch counts are shown in Figure 3.10a and b, respectively. The smaller particles, which were observed in sample C, had a high surface activity and specific surface area. As shown in Figure 3.10a, the residual particle counts of sample C are much larger than that of sample A. These residual particles induced the surface scratching during the CMP process. In Figure 3.10b, the scratch counts of sample C are also larger than sample A. Consequently, the ceria powders that include

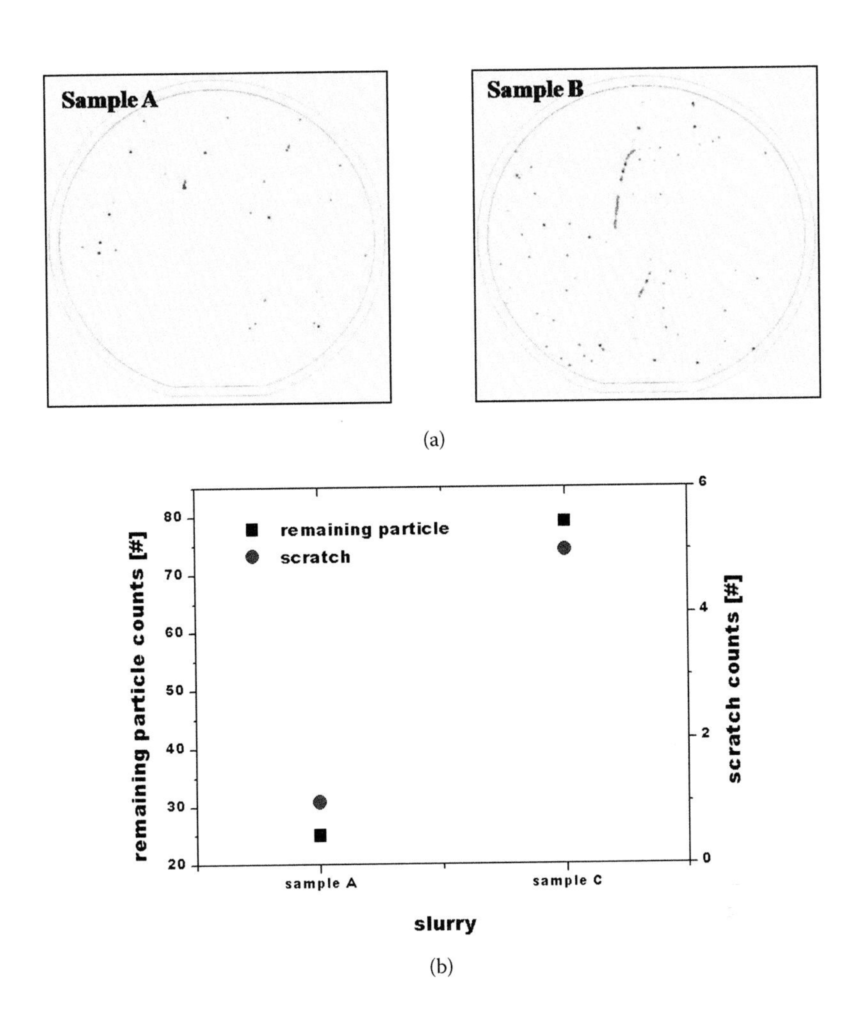

FIGURE 3.10 Remaining particle maps on the oxide film measured with the Sufscan 7700.

hexagonal structured particles were easily broken down to the smaller particles and induced particle adhesion on the wafer surface. Therefore, the ceria particles should be calcined to a cubic structure through control of the oxygen concentration.

3.3 Chemical Engineering for High Selectivity in STI CMP

STI CMP performance depends on several factors such as deposited nitride thickness, trench depth, thickness and type of deposited trench-fill oxide, removal rate consistency, the physicochemical nature of CeO_2 particles, organic additives, and suspension pH. Recently, many scientists have investigated the polishing mechanism of SiO_2 film and the physical properties of ceria particles. However, little is known about the removal rate and the selectivity between SiO_2 and Si_3N_4 CVD films. There are both chemical and mechanical interactions between ceria particles and the deposited film and the organic additives. The polishing of the oxide and nitride films is mainly affected by the nature of the CeO_2 abrasive particles and by some other chemical factors. The removal selectivity is due to the electrokinetic behaviors of the ceria particles and the SiO_2 and Si_3N_4 CVD films in aqueous media. The CMP performance can be influenced by differences in the surface potentials of ceria particles with poly(methacrylic acid) (PMAA), as dispersant, and the PETEOS and CVD Si_3N_4 films. The surface potentials are affected by the suspension pH. This results from the different isoelectric pH (pH_{iep}) of the ceria particles with PMAA, SiO_2, and Si_3N_4 CVD films.

3.3.1 Electrokinetic Behavior of the Ceria Particle, Oxide, and Nitride Films

The electrokinetic behaviors of ceria, ceria particles with PMAA, and SiO_2 and Si_3N_4 as a function of pH were investigated to identify the polishing behavior in STI CMP. These results are shown in Figure 3.10. The electrokinetic behavior of each particle is reflected in the interaction between the ceria particles in CMP slurry and the deposited film to be polished. The electrophoretic mobility of all components is strongly dependent on the suspension pH. The electrophoretic mobility of silica is negative above pH 3.4, which is the isoelectric point (pH_{iep}) of silica. However, at a pH above 9, the electrophoretic mobility decreases with an increasing suspension pH. This is attributed to a compression of the electrical double layer due to both dissolution of the Si^{4+} ion, resulting in an increase of ionic silicate species in the solution, and to the presence of some other alkaline ionic species. Si_3N_4 has a pH_{iep} of 6.5. Above this pH value, particles carry a net negative charge because of the formation of SiO^-, which results from the dissociation of surface silanol groups. The ceria particles have a pH_{iep} of about 7, and a slightly positively charged surface below this pH region. However, a shift in the pH_{iep} of the ceria particles toward the acidic pH is produced in the presence of PMAA. There are two causes for this behavior. First, the ionization of near-surface segments partially screens the charge on the particles, thereby decreasing the shear-plane potential. Second, the presence of polymer chains (PMAA) may disturb the hydrodynamic plane of shear, shifting it farther out from the particle surface. Because the potential decreases exponentially with the distance, the modified shear plane will experience a lower potential. Above the pH_{iep}, electrophoretic mobility of ceria particles increased with suspension pH up to the saturation point. Saturation occurs in the pH region of 6–9, resulting in a negatively charged particle whose electrokinetic behavior is essentially no longer dependent on the suspension pH. Thus, differences in the surface potentials of ceria particles with ionizable acrylic polymer, PETEOS, and silicon nitride films were found by electrokinetic analysis. The SiO_2 film and abrasive ceria particles are negatively charged above pH 4, but the Si_3N_4 film is positively charged below pH 7. These results cause the selective adsorption of acryl-based polymer. Additionally, the results suggest that the effect of dispersion stability on the polishing rate depends on the surface potentials between ceria particles and deposited films, which are influenced by the suspension pH. Ceria particles have negligible charges in acidic pH and thus agglomerate as a result of the weak interparticle repulsive forces. The low removal rate in this pH region is due to the decrease of contact area between ceria particles and the deposited film. Thus, the number of total particles participating in CMP decreases due to agglomeration. However, ceria particles have a negative charge in neutral and alkaline pH regions, where stable dispersion stability enables high removal rates of the PETEOS and CVD Si_3N_4 films.

Figure 3.11 shows the particle size distribution and the agglomerated particle behavior of ceria slurries for STI CMP as a function of suspension pH. Figure 3.11a shows that the acidic suspension had greater particle size and broader size distribution of abrasive ceria than those of the neutral or alkaline suspensions. The ceria slurry in acidic suspension was found to be unstable with a large agglomeration of particles, having a wide size distribution ranging from 0.02 to 100 μm resulting from lower electrostatic repulsive forces. The surface potential of ceria particles decreases in acidic suspensions. Figure 3.11b shows the agglomeration of abrasive ceria particles as a function of the suspension pH by in situ optical microscopy. A great number of large agglomerates were observed in the acidic slurry. The suspension became unstable because of a decrease in electrokinetic potential in the acidic pH region. The ceria slurries in neutral and alkaline suspensions were found to be well dispersed and large agglomerates were not observed.

3.3.2 STI CMP Performance in Different Suspension pH

The results of the CMP field evaluation are shown in Figure 3.12, and the removal rates of oxide and nitride and the selectivity of slurries are presented in Table 3.2. The slurries prepared in neutral and alkaline suspensions had high removal rates of PETEOS and low WIWNU.

As shown in Figure 3.10, the polishing rate is dependent on the electrokinetic behaviors among ceria particles and the PETEOS and CVD Si_3N_4 films at a given pH. This result was only considered from the interaction between the ceria particles. Furthermore, the removal rate is also affected by the suppression of Si_3N_4 film removal by the additive polymer. In this study, poly(acrylic acid) (PAA) was used to suppress the removal of the CVD Si_3N_4 film. The interactions between the PAA and the ceria particle/

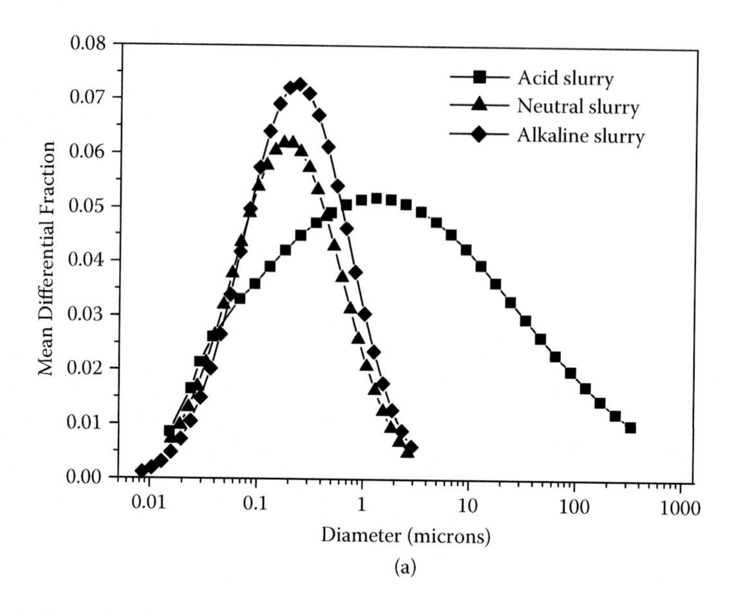

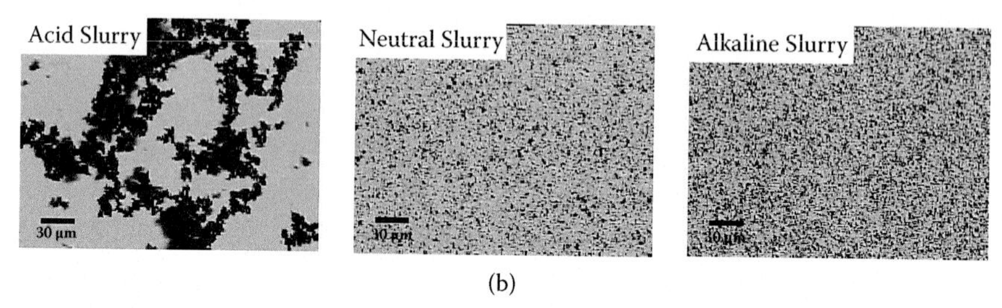

FIGURE 3.11 (a) Particle size distribution of ceria slurries; (b) optical micrographs of ceria slurries in suspensions of different pH.

TABLE 3.2

Removal Rate and Selectivity in CMP Field Evaluation

	Oxide R/R (Å/min)	Nitride R/R (Å/min)	Selectivity (Oxide R/R:Nitride R/R)
Acidic slurry	156.6	85.8	1.8:1
Neutral slurry	3098.9	69.7	44.5:1
Alkaline slurry	3718.7	731	5:1

deposited films were controlled by suspension pH. That is, coiled conformation of PAA at pH 3 permits a denser spacing on the PETEOS and CVD Si_3N_4 films, which results in a decrease in the removal of PETEOS and silicon nitride film. The PAA is adsorbed on both the PETEOS and CVD Si_3N_4 films, which suppresses the deposited film's removal due to formation of

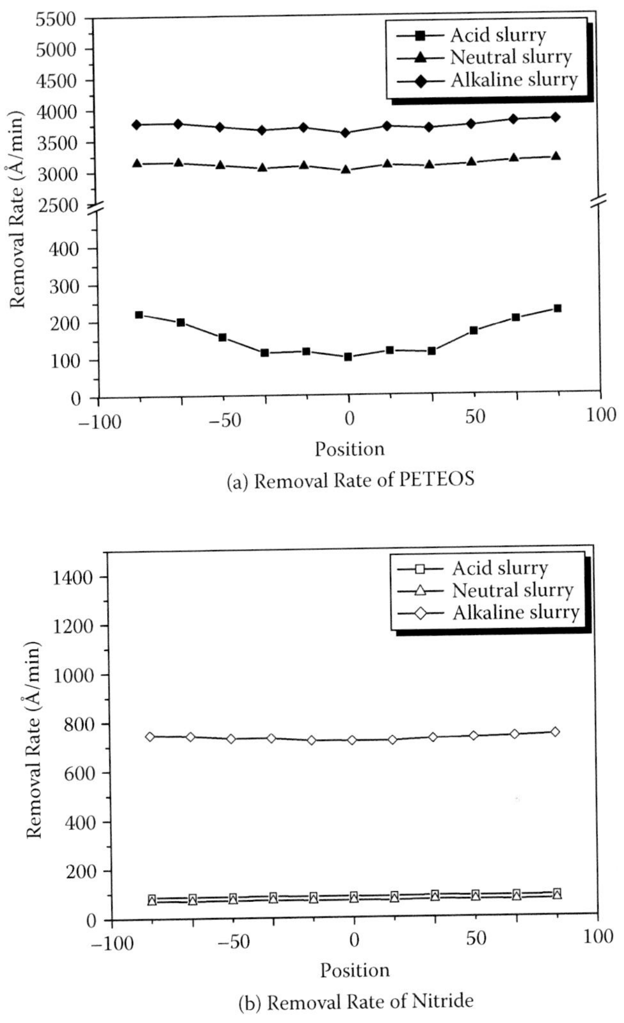

FIGURE 3.12 The result of CMP field evaluation: (a) average PETEOS removal rate and (b) average nitride removal rate.

a polymer-coated layer. As shown in Figure 3.11, dispersion stability of acidic slurry was found to be unstable, which resulted in a low PETEOS removal rate as well as worse non-uniformity in the CMP field evaluation. During CMP at pH 6, the ceria particles with PMAA, as dispersant, have a negative charge, and the PETEOS and CVD Si_3N_4 films have negative and positive charges, respectively. The PAA has negatively charged functional groups. The CVD Si_3N_4 film will have a higher affinity to this polymer than the PETEOS film would, that is, the PAA adsorbed only onto silicon nitride film with the formation of a polymer-coated layer. This was attributed to the selective adsorption due to the charge difference between the

PETEOS and CVD Si_3N_4 films in the neutral pH region. The suppression of silicon nitride film removal occurred due to the adsorption of PAA. Thus, high removal rates of PETEOS and oxide-to-nitride selectivity were obtained at pH 6. At pH 9, all materials have negative surface charges, resulting in a net negative charge, causing repulsion between the PAA and the surface of PETEOS and CVD Si_3N_4 films. Consequently, there is no polymer-coated layer formed on PETEOS and CVD Si_3N_4 films in an alkaline environment. The removal rate of the PETEOS and the CVD Si_3N_4 films both showed higher values at pH 9. In addition, Si^{4+} ions on PETEOS film dissolved at pH 9, which caused the PETEOS film to become softer. Therefore, the highest removal rate of the PETEOS was obtained in the alkaline pH region. Based on the results, high removal rates of PETEOS film and selectivity can be obtained by using neutral slurry.

The removal rate of the STI CMP process depends not only on the electrokinetic behaviors of abrasive ceria particles and the deposited films, but also on the conformation of the adsorbed polymer. The additive polymeric chains were found to have an opposite or identical charge with respect to the deposited film in certain pH regions. Thus, the selective adsorption of polymeric chains onto the surfaces of the CVD Si_3N_4 film in the neutral pH region was due to the CVD Si_3N_4 film having higher affinity than the PETEOS film for this polymer. Consequently, it is concluded that the control of suspension pH, which enables the dispersion stability and selective adsorption of polymeric chains, is the technical key for good uniformity of the deposited film surface and for high selectivity of STI CMP.

3.3.3 The Conformation of Polymeric Molecules and STI CMP Performance

Aqueous titration was carried out to observe the ionization behavior and conformational change of PAA as a function of pH. Figure 3.13 shows the degree of ionization, α, at different ionic strengths, where $\alpha = COO^-/(COO^- + COOH)$. As the ionic strength increased from 0 to 0.4 M, titration curves for PAA were slightly shifted to the left and the degrees of ionization of PAA increased from 0.743 to 0.825 at pH 6.5. The addition of KNO_3 provides counterions to stabilize the negative charges of carboxylate groups along the backbone of PAA, which subsequently shifts the equilibrium to the left. The counterions also form an ionic atmosphere around the negatively charged carboxylate groups of PAA, which shields the Coulombic interactions between COO^- and H^+ and screens the repulsion between the carboxylate groups. The reduced intersegment repulsion between screened carboxylate groups leads to a conformational change from a stretched to a coiled configuration.

Figure 3.14 shows the adsorption isotherms for PAA on silicon nitride as a function of ionic strength. By increasing the ionic strength from 0 to 0.4 M, the amount of PAA adsorbed on the silicon nitride surface increased

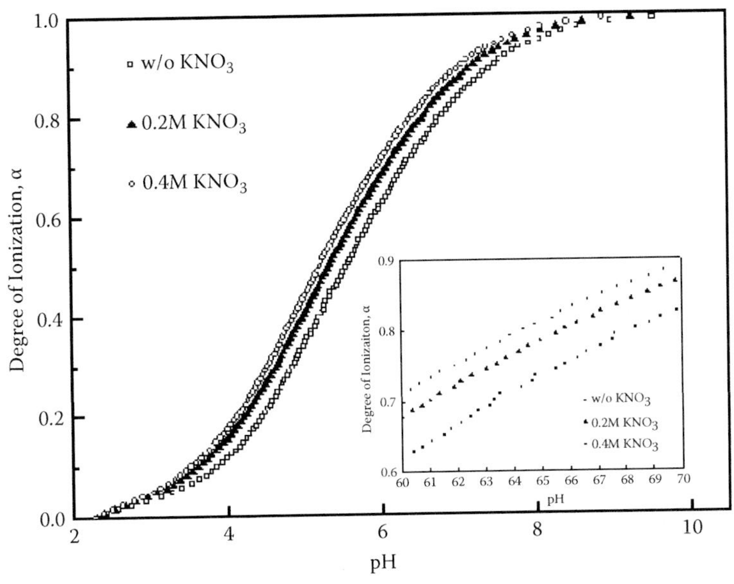

FIGURE 3.13 Dissociation of PAA as a function of pH and ionic strength.

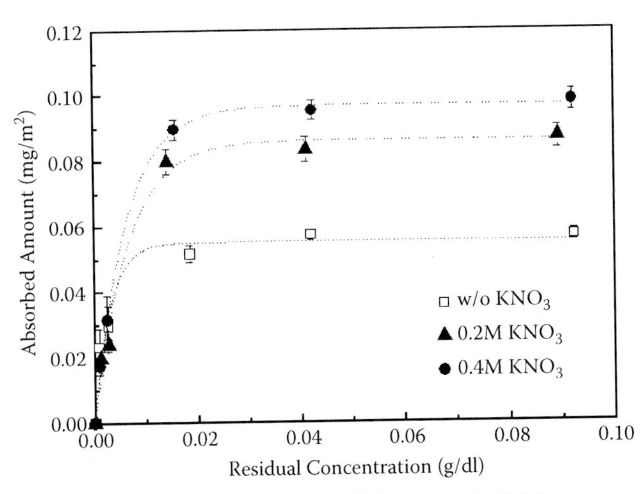

FIGURE 3.14 Adsorption isotherms for PAA on SiN surface at pH 6.5.

from approximately 0.055 to 0.097 mg/m^2. The amount of PAA adsorbed in the presence of KNO_3 was increased by the charge neutralization of PAA due to the screening of the electrostatic attraction between COO^- and H^+. The dissolved potassium ions can form an ionic atmosphere around the negatively charged carboxylate groups of PAA, and it screens the electrostatic repulsive forces between adjacent carboxylate groups. This screening results in the charge neutralization of negatively charged carboxylate groups. Thus the conformation of PAA was known to change from a stretched to a coiled structure by charge neutralization. The conformation of the polymeric molecules affected the adsorption behavior of polymeric molecules on the solid phase, which is primarily affected by the suspension pH, the molar mass of polymeric molecules, and the number density and valence of the counterions. A low ionic strength permits a stronger interaction with the aqueous phase and reduces the driving force for adsorption onto the Si_3N_4 due to the electrostatic repulsion between neighboring carboxylate groups in the PAA chains. This results in a reduced uptake and a more extended adsorption conformation with fewer attachment points on the Si_3N_4 surface. In contrast, a high ionic strength results in a stronger, more compact interaction with the Si_3N_4 surface through nonelectrostatic interactions between the Si_3N_4 surface and PAA chains.

Figure 3.15 represents the effect of ionic strength on the intersegment forces between the adsorbed PAA on Si_3N_4 film and Si_3N_4 AFM tip. The strong repulsive force of the PAA solution in the absence of KNO_3 originates at a separation of 30 nm. This was attributed to the electrostatic repulsion between the negatively charged carboxylate groups of PAA at a pH 6.5. The separation distance was decreased by increasing concentrations of KNO_3. With the addition of 0.4 M KNO_3, the repulsive force was observed at the reduced separation distance of 5 nm. This result confirms that charge neutralization for the negatively charged carboxylate group of PAA is accelerated when the ionic strength is higher. In particular, as can be seen in the logarithmic plot in the inset of Figure 3.15, the repulsive forces of the PAA solution with 0 M, 0.2 M, and 0.4 M KNO_3

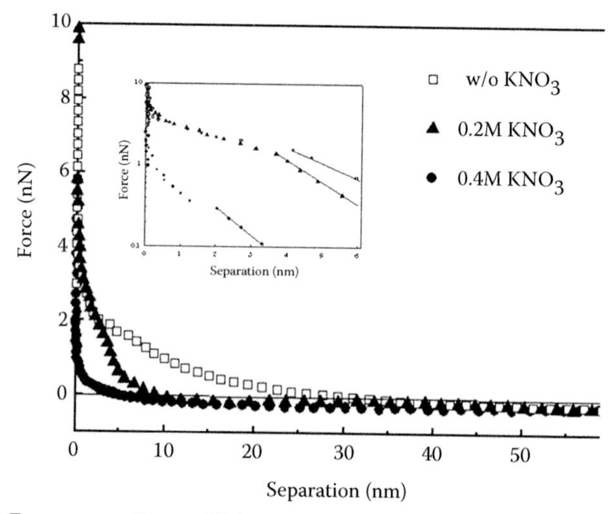

FIGURE 3.15 Force–separation profile between the AFM tip and a SiN film with different concentration of KNO_3 at pH 6.5.

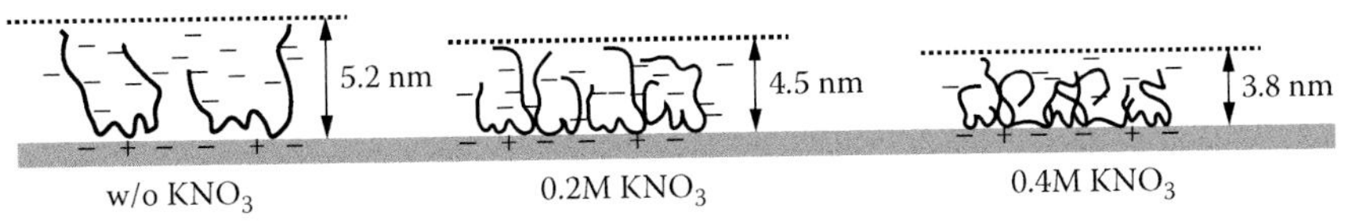

FIGURE 3.16 Schematic diagram of the adsorption behavior and conformation of PAA on SiN film.

follow the Poisson–Boltzmann equation under the separation distances of 4.2, 3.6, and 2.0 nm, respectively. Below these separation distances, the repulsive force between the PAA adlayers is dominated by steric hindrance. Therefore, the adsorption thickness of PAA on the nitride film can be considered to be decreased from 2.1 to 1.0 nm by increasing the concentration of KNO_3 to 0.4 M. With the increase of ionic strength, the thickness of PAA adsorbed on the Si_3N_4 decreased. The thicknesses of the PAA adlayer on the Si_3N_4 surface was 5.2, 4.5, and 3.8 nm for PAA with 0 M, 0.2 M and 0.4 M KNO_3, respectively. Atomic force microscopy (AFM) and Nanospec data show a similar trend; the thickness of PAA decreases as ionic strength increases. This implies that the coiled conformation of PAA forms a dense passivation layer on the Si_3N_4 film. The adsorption behavior and conformational change of PAA are depicted schematically in Figure 3.16. The trends in removal rates along the whole of the wafer for the SiO_2 and Si_3N_4 layers are shown in Figure 3.17. Since the PAA solution with the addition of 0.4 M KNO_3 is expected to effectively suppress the removal of the Si_3N_4 films, a CMP field evaluation was performed using a PAA solution without KNO_3 and with 0.4 M KNO_3. The removal rate of the SiO_2 film was 2690 Å/min, 2657 Å/min and 2610 Å/min for PAA with 0 M, 0.2 M, and 0.4 M KNO_3, respectively. As the ionic strength increases, the particle is agglomerated by screening the charges on the particle surfaces, which results in increasing the removal rate. However, for the purposes of the present study, the KNO_3 was added into the PAA solution, then the ceria slurry and the PAA solution are mixed at the point of use (POU). Therefore, the removal rate of the SiO_2 film was affected by, not the ionic strength, but the size and crystallinity of the ceria particle. However, the PAA solution with 0.4 M KNO_3 resulted in lower removal rates of the CVD Si_3N_4. The average removal rate of the Si_3N_4 film decreased from 72 Å/min to 61 Å/min as ionic strength increased up to 0.4 M. The characteristics of the CMP evaluation are attributed to the conformational change of PAA. As ionic strength increases, negatively charged carboxylate groups of PAA are neutralized by the presence of potassium ions and thus PAA adopts a coiled configuration. This leads to a reduction in the potential for the penetration of abrasive particles into the film and decreases the friction force between the abrasive particles and the film due to the dense passivation layer. As a result, the Si_3N_4 film was successfully passivated by using a PAA solution with 0.4 M KNO_3, which resulted in a lower removal rate than the PAA solution without KNO_3. Consequently, the SiO_2-to-Si_3N_4 selectivity increased from 37:1 to 42:1 as ionic strength increased up to 0.4 M. It is noteworthy that the successful passivation of Si_3N_4 in STI CMP can be accomplished by controlling the conformation and adsorption amount of PAA.

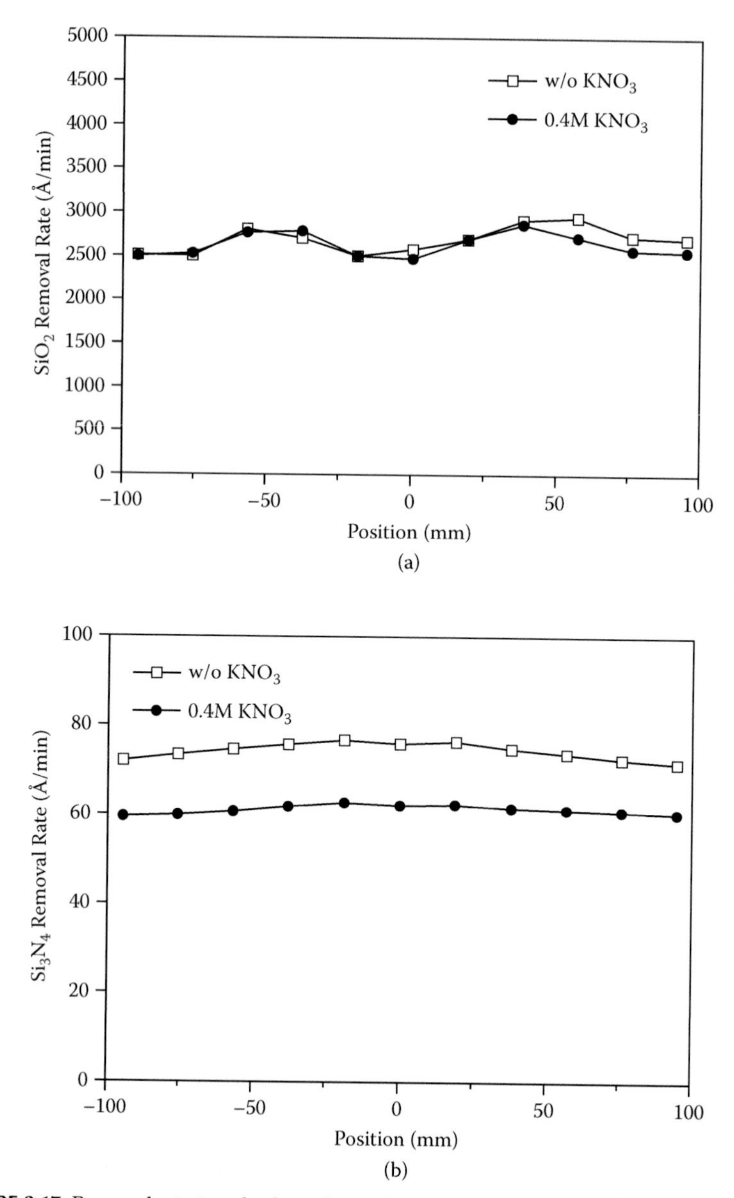

FIGURE 3.17 Removal rate trends along the entire resultant (a) SiO_2 and (b) SiN film.

3.4 Force Measurement Using Atomic Force Microscopy for Mechanism

The force–separation profile between the AFM tip and the PETEOS or CVD Si_3N_4 film was measured to analyze the interaction between PAA and the deposited film (Figure 3.18). Force measurement using AFM is expected to indicate the direct evaluation of the adsorption behavior of the polymeric molecules on the film surface. Figure 3.18a shows the force–separation plot of the Si tip with PETEOS film at pH 6.5 for different molecular weight of PAA. The surface of the oxide film, whose pH_{iep} is about 3.4, is negatively charged at pH 6.5, and thus a repulsive force occurred due to the electrostatic interaction. The interaction range is about 20 nm in the absence of PAA. It is of interest that there is no significant difference between the surface forces of the tip and the oxide film even with the presence of PAA. This result is almost the same for all samples, irrespective of molecular weight, which means that PAA is scarcely adsorbed on the oxide film due to the electrostatic repulsion between like-charge of the film and PAA. This is because the adsorption behavior of partially or fully ionized polyelectrolyte such as PAA is dominantly affected by electrostatic interaction.

On the other hand, Figure 3.18b illustrates a meaningful change in the interaction force between the nitride film and Si_3N_4 tip with and without PAA having the different molecular weight of 5,000, 15,000, and 30,000. In the absence of the absorbed PAA molecule, an attractive force was observed at approximately 10 nm of separation distance and the maximum attractive separation force is about 0.5 nN at 5 nm of separation. However, it was found that the attractive force disappeared and a repulsive force was shown by the addition of PAA. This result is more clearly observed as the molecular weight of PAA increases. In PAA 5,000 solution, it was shown that the attractive force drastically decreased, even though a very weak attraction still exists.

Strong repulsive forces were observed for PAA 15,000 and 30,000 solutions. In the case of PAA 15,000 and 30,000, the repulsion force starts to originate at a separation of 30 nm. Especially, a more significant increase in the interaction force is observed from about separation of 7 nm in the case of PAA 30,000 solution, compared to PAA 15,000 solution. The pH 6.5, where the experiment was performed, is almost the same as the pH_{iep} of silicon nitride, and thus net charge forces on the surface of both the tip and wafer is nearly zero in the absence of PAA, which results in a van der Waals attractive force. However, the presence of PAA in the system leads to the adsorption of PAA on the nitride film and the resulting formation of PAA layer, and thus the repulsion between the dense PAA layers adsorbed onto the tip and the film is observed. As can be seen in

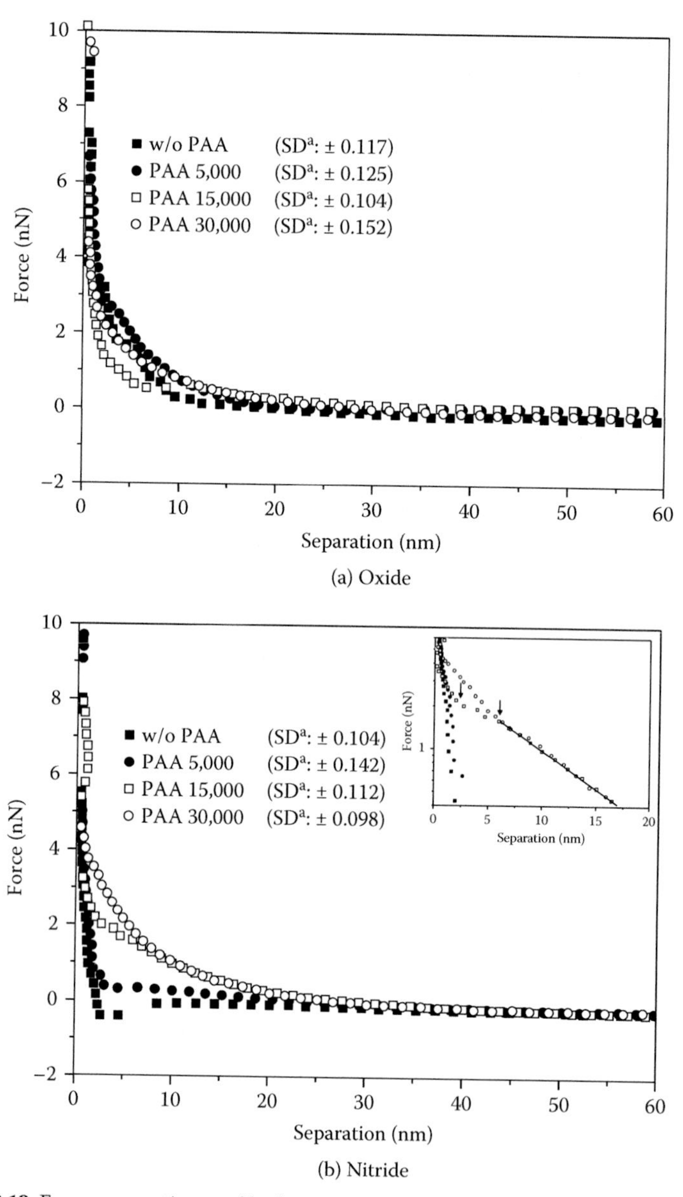

FIGURE 3.18 Force–separation profiles between the AFM tip and a film with different Mw of PAA at pH 6.5: (a) oxide, (b) nitride.

the logarithmic plot in the inset of Figure 3.18b, the repulsive forces of PAA 15,000 and PAA 30,000 follows the Poisson–Boltzmann equation up for 4 and 7 nm of the separation distance, respectively, which indicates that the interaction between PAA layers are of electrostatic origin. Below these separation distances, the PAA layer is responsible for the repulsive interaction, that is, the steric contribution. It is suggested that the adsorption thickness of PAA on the nitride film is 2 nm for PAA 15,000 and 3.5 nm for PAA 30,000. Thus, the increase in repulsion force with molecular weight of PAA is attributed to the increase of the PAA adsorption layer thickness and the formation of a denser adsorption layer. This AFM result agrees with Vedula and Spencer's work. They revealed that the maximum adsorption amount of PAA is increased as the molecular weight increases and the adsorption behavior of PAA on amphoteric surfaces in an aqueous suspension is qualitatively similar for a variety of materials, including Si_3N_4.

The different adsorption behavior of PAA on the PETEOS/CVD Si_3N_4 can be further described by the AFM images (Figure 3.19). The morphology and surface roughness of the oxide and nitride film with and without PAA 30,000 were observed. In the case of oxide, there is no significant change of the film surface between the bare oxide film and the PAA 30,000 spin-coated film, which means PAA is not adsorbed on the oxide film, as shown in Figure 3.19a and Figure 3.19b. On the other hand, it was clearly found that the surface roughness of the PAA 30,000 spin-coated nitride film (0.469 nm) is much higher, compared to that of the bare nitride film (0.156 nm). This is attributed to the formation of the PAA adsorption layer on the nitride film due to the electrostatic interaction between PAA and the film surface. These results are in agreement with the AFM force measurement shown in Figure 3.18b.

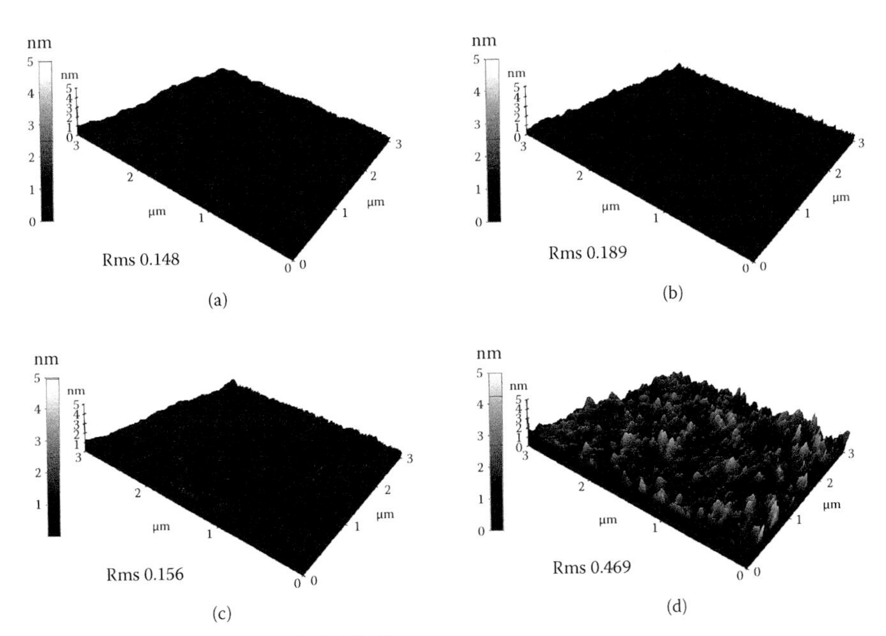

FIGURE 3.19 AFM images of nitride film.

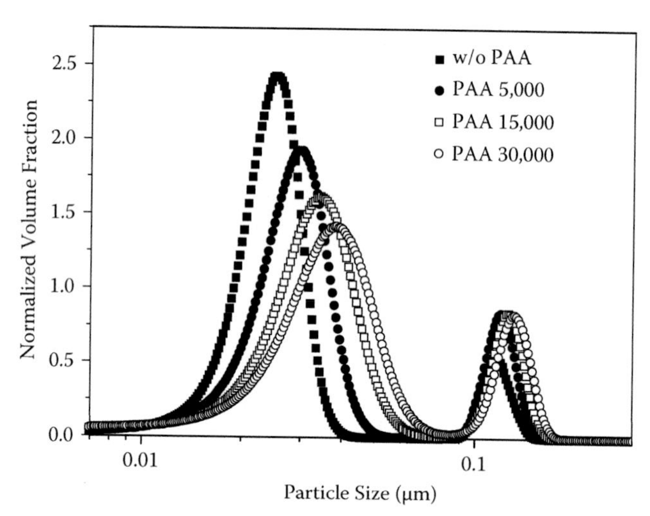

FIGURE 3.20 Particle size distributions with different Mw of PAA.

Additionally, it is reported that PAA in the CeO_2 slurry has an influence on the dispersion stability of the CeO_2 particles. Because the suspension stability is significantly reflected in the interaction between the CeO_2 particles in the CMP slurry and the deposited film, the relationship between the organic additive having the different molecular weight and the dispersion stability of the CeO_2 particles should be considered. Figure 3.20 illustrates the particle size distribution of CeO_2 as a function of the molecular weight of PAA. It was found that the size distribution of CeO_2 becomes broader and the mean secondary particle size is increased from 37 nm to 46 nm as the molecular weight of PAA is increased from 0 to 30,000. Also, the volume fraction of the agglomerated particles increases with an increase in molecular weight of PAA, indicating that particle agglomeration occurred in shallow secondary minima with a decreasing of electrical repulsive forces. Therefore, it is expected that the increase in molecular weight of PAA leads to the decrease in the suspension stability.

The results of the CMP performance as a function of the molecular weight of PAA are shown in Figure 3.21. It was found that WIWNU is maintained at a low level, regardless of the molecular weight of PAA. This indicates that all of CeO_2 slurries used in this study have reasonably good dispersion stability, even though there is a slight difference in the agglomeration particle size among them. The removal rates of both the oxide and nitride films are decreased as the molecular weight of PAA increases. The removal rate of the average oxide film proportionally decreases from 3365 Å/min to 2420 Å/min as the molecular weight of PAA increases from 0 to 30,000. This result is attributed to the relationship between the stability of the CeO_2 slurry and the molecular weight of PAA as shown above. A stable dispersion enables a high removal rate of the deposited film due to an increase in the contact area between the particles and the film because the number of total particles

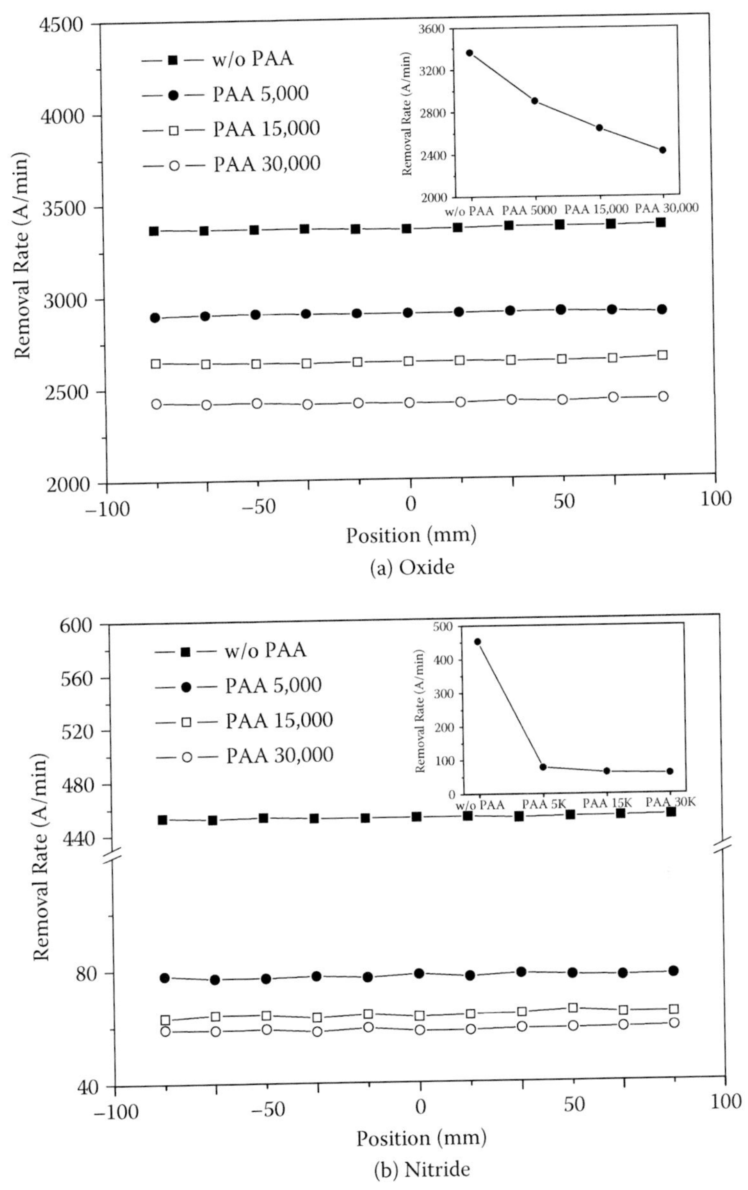

FIGURE 3.21 Removal rate trends along the entire resultant (a) oxide and (b) nitride film.

participating in the CMP process increases. However, it is noteworthy that the removal rate of the nitride dramatically decreases from 453 Å/min to 59 Å/min with increase in molecular weight of PAA from 0 to 30,000, indicating that it does not show a linear relationship with the dispersion stability of CeO_2.

From the AFM result shown in Figure 3.18, it can be considered that this drastic decrease in the removal rate of the nitride film as a function of the molecular weight of PAA is due to the formation of the passivation layer due to the strong interaction between PAA and nitride film. Force measurement using AFM clearly showed that the adsorption behavior of PAA, namely, the adsorption layer thickness on the nitride film, is strengthened with an increase in molecular weight. In addition, the PAA layer formed on the nitride film is scarcely removed, but the PAA layer can be sustained under the mechanical stress of the down pressure and spindle force during the STI CMP process.

This phenomenon can be explained by two mechanisms: (1) the strong adhesion occurs due to the electrostatic interaction between PAA and the nitride film at pH 6.5 and (2) the adsorbed or nonadsorbed PAA molecule reduces the mechanical stress on the nitride film, acting as lubricant. Although the destruction (delamination) of the PAA layer on the nitride film occurs under an applied load, the existence of PAA in the bulk solution can be dynamically readsorbed on the nitride surface, which is newly exposed after the removal of PAA layer. Thus, it leads to the reduction of the possibility of the penetration of the abrasive CeO_2 particles on the wafer and the decrease of the friction force between the abrasive particle and nitride. On the other hand, PAA is seldom adsorbed on the oxide film, regardless of molecular weight, due to the electrostatic repulsion between the like-charge of the PAA and the oxide film surface. Therefore, this difference in the removal mechanism of the oxide and nitride films gives rise to the dramatic increase in the oxide-to-nitride selectivity values from 7:1 to 41:1 as the molecular weight of PAA increases from 0 to 30,000. Thus, the formation of a passivation layer on the nitride film by the adsorption behavior of the polymer plays a dominant role in determining the oxide-to-nitride selectivity in STI CMP, and this selectivity is significantly affected by the molecular weight of the polymer. Consequently, it was obtained by the force measurement using AFM that the control of the molecular weight of PAA is a technical key to obtaining high oxide-to-nitride selectivity in STI CMP.

3.5 Pattern Dependence of High-Selectivity Slurry

To improve the performance of high-selectivity ceria slurry in STI CMP, it is essential to control the slurry properties, including the pH, the concentration, the molecular weights of the organic additives, and the abrasive particle size. The dependencies of the removed amount and the surface roughness of SiO_2 and Si_3N_4 films on the molecular weight and the concentration of PAA in ceria slurries containing abrasives with different primary sizes, through STI CMP tests using blank and patterned wafers, were investigated.

Cerium carbonate was used as a precursor to synthesize two types of ceria powder. The primary grain size of the polycrystalline ceria abrasives was controlled by employing a calcination process for 4 h with two calcination temperatures of 700°C and 800°C. The secondary particle size of the abrasives was controlled by crushing the powders by using a laboratory-scale air jet mill and a wet ball mill. The ceria powders were crushed by wet mechanical milling for several hours to reduce their secondary particle sizes to the target size of 130 nm, after initial mechanical dry jet milling for several hours to reduce the size to 300 nm. The ceria abrasives were then dispersed in deionized water and stabilized by adding 100 ppm of a commercially available dispersant (PMAA), along with 1 wt% of ammonium salt (Mw = 10,000; Darvan C, R.T. Vanderbilt, USA) as another dispersant of the abrasive particles. We also added an anionic organic additive (PAA; Polysciences, USA) at a concentration of up to 0.80 wt%, with one of three molecular weights (Mw = 30,000, 50,000, and 90,000). Each suspension was twice subjected to ultrasonic treatment for 15 min to break down agglomerates and promote mixing. An ice bath was used to control the temperature of the suspension during the ultrasonic treatment. The suspension was aged for 12 h at room temperature with a wrist-action shaker and subjected to ultrasonic treatment for an additional 15 min prior to use. The solid content was initially controlled to 5 wt% of ceria powder in the suspension. We then diluted each slurry with deionized water to produce a final ceria abrasive concentration of 1 wt%. Each slurry's pH was adjusted to the range of 6.0 to 7.0 by adding an alkaline agent. Table 3.3 lists the slurry characteristics, including the slurry pH, the different PAA pH values with the three molecular weights, and the experimental conditions during synthesis.

The crystal structure and grain size were analyzed with a diffractometer (RINT/DMAX-2500, Rigaku, Japan) using Cu–Kα radiation ($\lambda = 0.1542$

TABLE 3.3
Summary of Slurry Characteristics

<table>
<tr><th rowspan="2">Sample</th><th rowspan="2">Slurry pH</th><th rowspan="2">Calcination Temperature (°C)</th><th colspan="2">Milling Time (Hours)</th><th colspan="3">Chemical Additive Characteristics</th></tr>
<tr><th>Dry</th><th>Wet</th><th>Concentration (wt%)</th><th>Molecular Weight</th><th>pH</th></tr>
<tr><td>A</td><td rowspan="4">9.0</td><td rowspan="3">800°C</td><td rowspan="3">4</td><td rowspan="3">35</td><td rowspan="4">0 ~ 0.57</td><td>30,000</td><td rowspan="4">6.5</td></tr>
<tr><td>B</td><td>50,000</td></tr>
<tr><td>C</td><td>90,000</td></tr>
<tr><td>D</td><td>700°C</td><td>2</td><td>32</td><td>30,000</td></tr>
</table>

nm) at a scan rate of 2° min^{-1} (2θ min^{-1}). The intensity was logged over a 2θ range of 25°C ~ 60°C with a scan step of 0.02°C. The grain sizes of the calcined powders were estimated using an x-ray line broadening method by applying the Debye–Scherrer equation. The morphology of the abrasives was observed with a high-resolution transmission electron microscope (HRTEM; JEM-2010, JEOL, Japan). The secondary particle size in each slurry was measured by acoustic attenuation spectroscopy (APS-100, Matec Applied Sciences, USA). Each suspension pH was measured with an advanced benchtop pH meter (Orion-525A, Thermo Orion, USA) by adding KOH and HCl to control the range of 7.0 to 8.0. The rheological behavior of the slurry suspensions was examined with a controlled-stress viscometer (MCR300, Paar Physica, Germany). This viscometer has concentric-cylinder geometry, enabling us to investigate the stability behavior of the slurry with an external-temperature-control bath circulator operating at 25 ± 0.1°C .

For blanket wafer tests, we used conventional 8-inch silicon wafers prepared by the single-side polishing method. SiO_2 films were deposited by the PETEOS method. Si_3N_4 films were formed by low-pressure chemical vapor deposition (LPCVD). For the patterned case, the SKW-3 pattern wafer designed by SKW Associates was used for characterization with respect to the pattern density and pitch size. The STI mask consisted of 4 mm × 4 mm density and pitch structures dividing the 20 mm × 20 mm die into five rows and five columns. Figure 3.22 illustrates the specially designed layout of the SKW-3 pattern wafer, including (a) the pattern density and pitch size layout, (b) the mask floor plan, and (c) a cross-sectional view. The thicknesses of the as-deposited SiO_2 and Si_3N_4 films on the blanket and SKW-3 patterned wafers were 700 nm and 150 nm, respectively. In the density structure (where density is defined as "trench width (TW)/ [trench width (TW) + active width (AW)]" or the trench area over the total area), the pattern density is varied systematically from 0% to 100% in increments of 10%, with a fixed pitch of 100 μm. The density structures are fabricated in a random layout in order to place high-density regions next to low-density regions. In the pitch structure, the density is fixed with the same trench width and space (50%), and the pitch is varied from 1 to 1000 μm, with vertically oriented lines. A cross-sectional pattern image of active Si_3N_4 and field SiO_2 layers before and after polishing were observed by scanning electron microscopy.

For the CMP process, the films were polished on a Strasbaugh 6EC polisher, with an IC1000/Suba IV stacked pad (Rodel, USA). The polishing pressure, applied as a down force, was 4 psi, equivalent to 27.5 kPa. The relative velocity between the pad and the wafer was 0.539 m/s. The polishing time was 30 s. The SiO_2 and Si_3N_4 film thickness variations of the wafers before and after CMP were measured with a NanoSpec 180 (Nanometrics, Milpitas, California) and an Opti-probe (Therma-Wave, Fremont, California). Cross-sectional images of the SKW-3 patterned

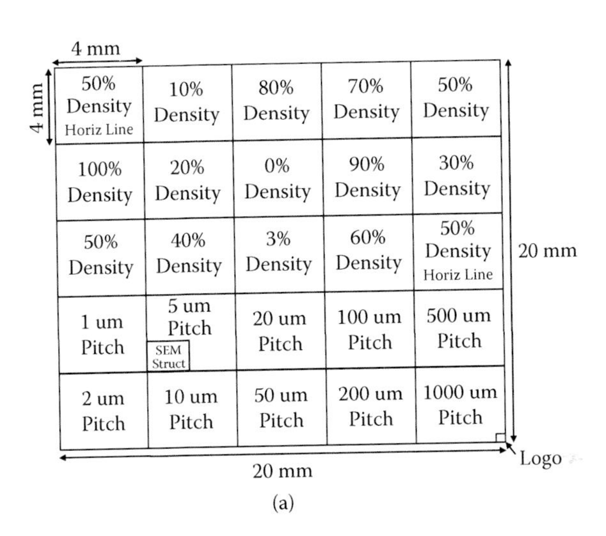

(a)

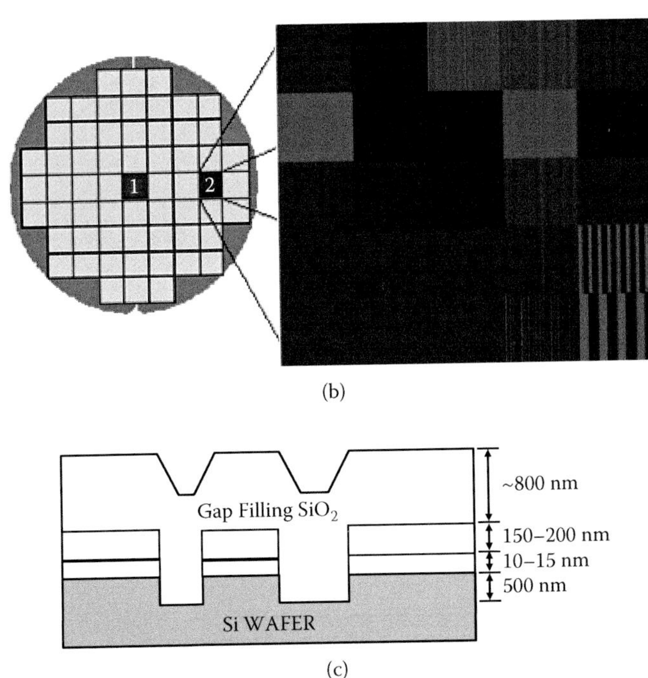

(b)

(c)

FIGURE 3.22 Specially designed layout of the SKW-3 pattern wafer: (a) pattern density and pitch size layout, (b) mask floor plan, and (c) cross-sectional view.

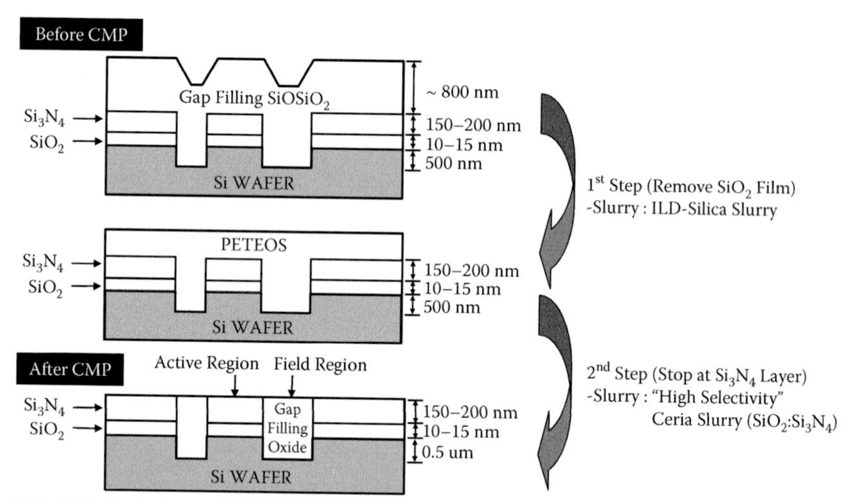

FIGURE 3.23 Schematic process flow of a typical STI CMP process.

wafers were obtained with a high-resolution scanning electron microscope (HRSEM; EP-1040, Hitachi, Japan). To analyze the surface roughness of the SiO_2 and Si_3N_4 films, an area of 1.0 μm × 1.0 μm was characterized with a commercial multimode atomic force microscope (AFM; XE 150, PSIA, Korea). A typical shallow trench structure was used to isolate the active regions where devices would be fabricated. The Si_3N_4 layer was attended, and a shallow trench was etched into the silicon, as illustrated in Figure 3.23. A SiO_2 film was then deposited into the trench, resulting in an overburden of SiO_2 above the Si_3N_4 active areas. In the ideal STI CMP process in mass production, the SiO_2 film is roughly removed in all local step-coverage regions, leaving SiO_2 film only in the trench regions. Fumed silica slurry was used for the first CMP step, while the ceria slurries were used for the second CMP step in the polishing process to achieve stopping on the Si_3N_4 film surface after complete removal of the SiO_2 film.

Figure 3.24 shows the HRTEM images and XRD powder diffraction patterns of the abrasive particles calcined at two temperatures (700°C and 800°C). These images indicated that the primary grain size increased with calcination temperature, and that the morphology of the ceria particle varied according to the calcination temperature. The abrasives calcined at 700°C showed a relatively low crystallinity, whereas those calcined at 800°C exhibited a relatively high crystallinity and the shapes of the grain are well defined, though some grains seem to contain subgrain boundaries inside. In the ring-shaped diffraction, the particles calcined at temperatures as low as 700°C still maintain their crystallinity as shown in Figure 3.24a. This result coincides with the XRD peaks shown in Figure 3.24b. In addition, the slurry calcined at 700°C contained both medium-sized particles and many small primary particles, whereas the

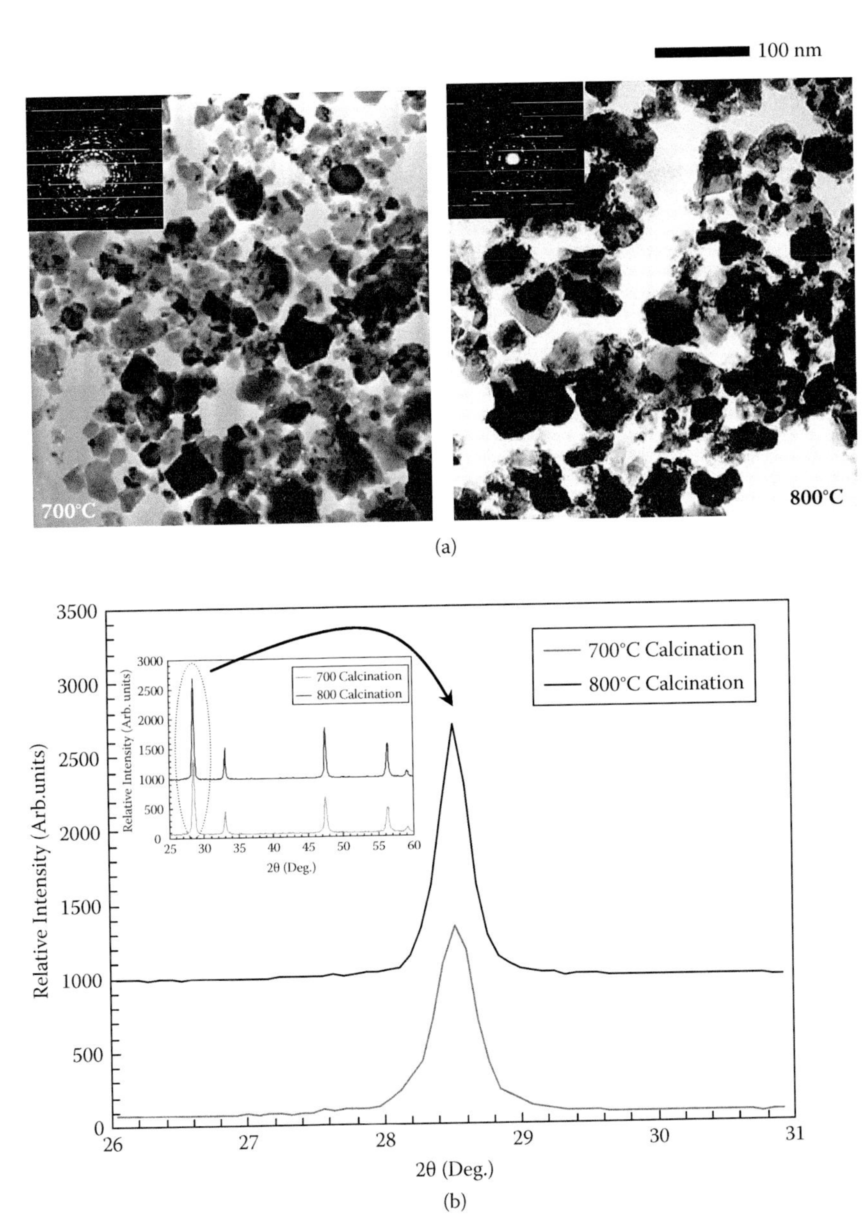

FIGURE 3.24 HRTEM photograph and XRD powder diffraction patterns of the abrasive particles in a ceria slurry: (a) HRTEM photograph and diffraction pattern and (b) XRD powder diffraction pattern as a function of calcination.

other slurry was composed of uniformly distributed, medium-sized particles. As confirmed by the TEM images, the slurry calcined at 700°C had a wider size distribution than that of the other slurry. The XRD patterns of the powders calcined at different temperatures is shown in Figure 3.24b. Broader intensity peaks were observed for the ceria powders, which were synthesized at 700°C. This result may be considered by the low crystallinity with unreacted cerium carbonate and small-sized abrasives grain.

The diffraction pattern only shows the peaks of cerium oxide with a fluorite structure; those for other compounds, such as cerium carbonate and cerous oxide, were not detected. With increased calcination temperature, the characteristic peaks of CeO_2 became sharper because the grains of single crystals were proportionally grown by heat treatment. This result affected the average grain size of the particles. The primary grain size of CeO_2 was investigated to clarify the relationship between the calcination temperature and the physical characteristics of the particles. The line broadening of the (111) peak in XRD was analyzed to confirm the primary grain size of particles. The intensity peak at $2\theta = 28.2°C$ was chosen for calculating the grain size, since it was clearer than any other peak and isolated from the others. The grain size moderately increased overall from 27 to 36 nm as the calcination temperature was increased from 700°C to 800°C, which can be attributed to thermally promoted grain growth during the calcination process. These results are in agreement with the trend of increasing grain size in the TEM images shown in Figure 3.24a.

Figure 3.25a shows the distributions of the secondary particle sizes for both slurries without PAA addition. There was no difference in the distribution for small particle sizes of 0 to 0.6 µm. On the other hand, the slurry calcined at 700°C had a distribution with a higher range of large particles (>3 µm) than the other slurry. Figure 3.25b shows the median sizes (d_{50}) of the abrasives in each slurry as a function of the PAA concentration. With increasing PAA concentration, the average secondary particle size gradually increased within the concentration range from 0 to 0.60 wt%.

The average secondary particle size of the polycrystalline abrasives in ceria slurry is thought to be determined predominantly by PAA adsorption on the abrasives particle in the ceria slurry suspension. Generally, the amount of anionic PAA adsorbed on the abrasive particle surfaces, the configuration of the adsorbed PAA molecules, and the electric surface charge adsorbed from the particles by the PAA polymer chains control the agglomeration state and the stability of the dispersion.

To evaluate the effects of the primary size of the ceria abrasives and the PAA concentration with different molecular weights on STI CMP, we conducted blanket wafer tests and measured the removal rates of SiO_2 and Si_3N_4 for the three slurry samples. Figure 3.26 show the results obtained from matrix experiment conducted by varying the molecular weights and concentrations of the PAA, along with the primary size of the ceria abrasives in each slurry. The removal rate of SiO_2 was reduced with increasing

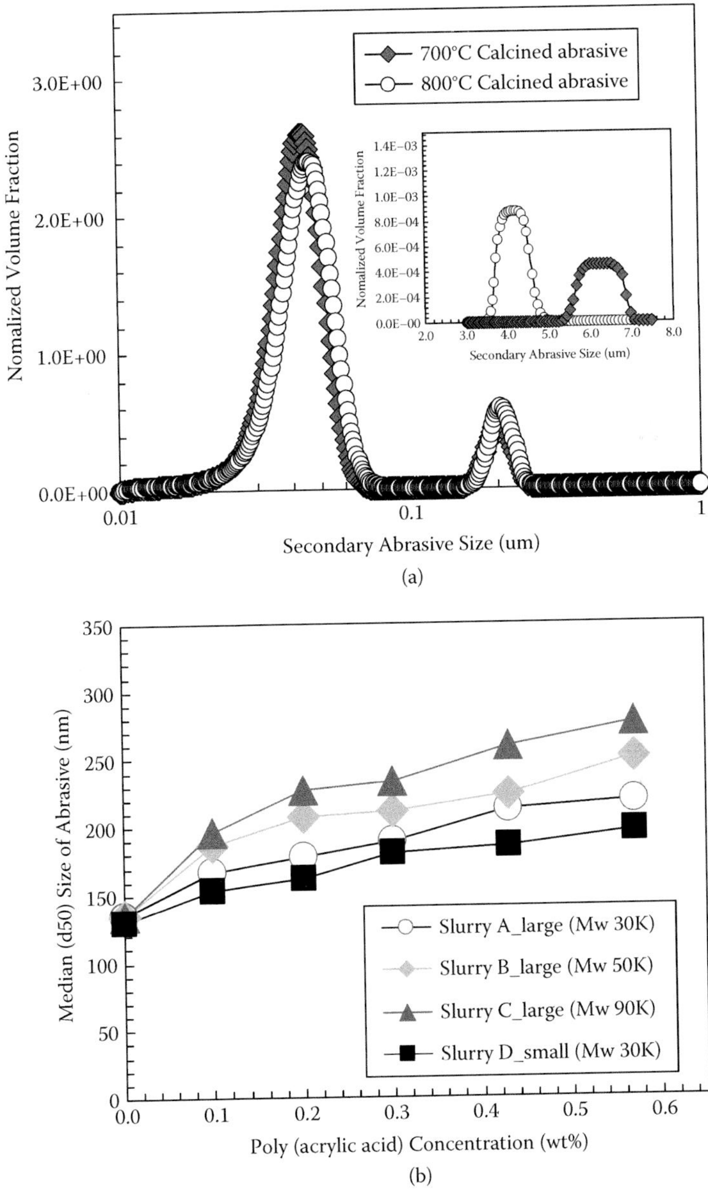

FIGURE 3.25 (a) Abrasive particle size distribution without surfactant addition and (b) average median (d_{50}) abrasive size as a function of the surfactant molecular weight at pH 6.5 to 7.0.

molecular weight for the same primary size throughout the experimental range of PAA concentrations, as shown in Figure 3.26a. For the PAA with the highest molecular weight with a different primary size, however, the removal rate of SiO_2 film was markedly reduced, from 2184 to 537 Å/min, as the PAA concentration increased. In contrast, in the case of the PAA

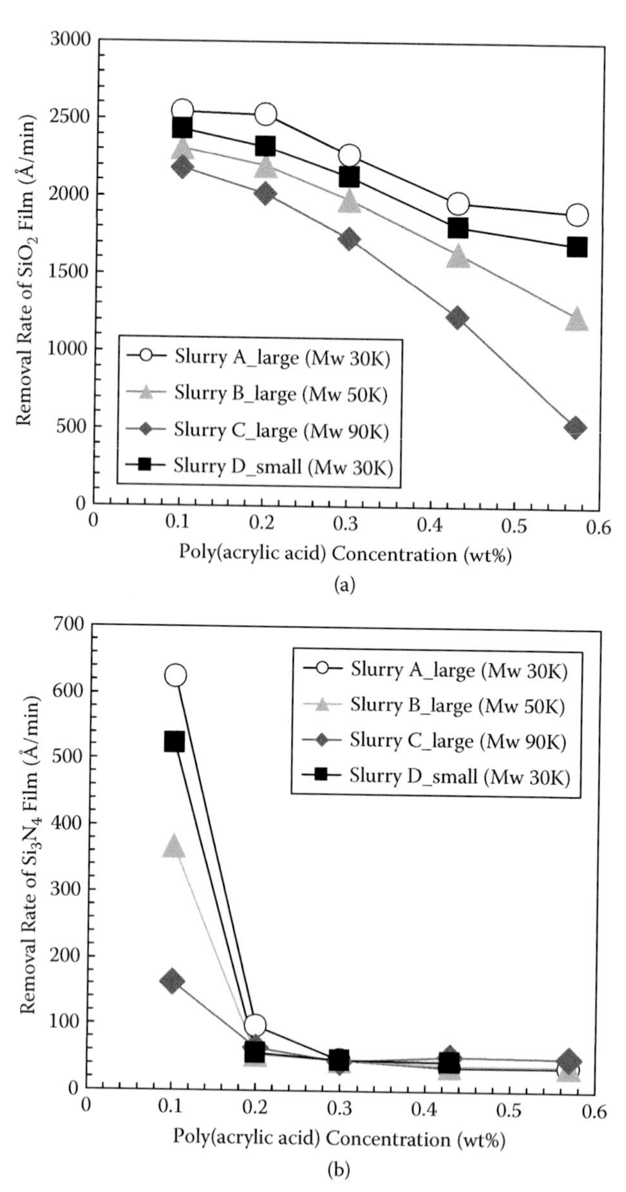

FIGURE 3.26 Results of the CMP tests of blanket wafers in terms of the surfactant molecular weight: (a) removal rate of SiO_2 film and (b) removal rate of Si_3N_4 film.

with the lowest molecular weight, the removal rate only slightly reduced, from 2542 to 1901 Å/min. Hence, with increasing PAA concentration, a higher primary abrasive size maintained a higher removal rate of SiO_2 at the same molecular weight. The removal rate of Si_3N_4 film versus the PAA concentration for slurries with the three PAA molecular weights and the two primary abrasive sizes is shown in Figure 3.26b. Kang et al. (2004) previously reported that the contact probability of the abrasives on the film surface should strongly influence the removal rate. The passivation layer is formed by PAA adsorbed on the film surface during CMP, and that the effectiveness of this layer may depend on the amount of selective adsorption on the film surface and on the concentration of PAA with increased molecular weight. Furthermore, we have attributed this to the behavior of abrasives moving in the PAA adsorption layer near the film surface. The removal rate of Si_3N_4 film was markedly reduced with increasing molecular weight, and it essentially saturated beyond a PAA concentration of 0.30 wt%. In addition, as a result of increasing the PAA concentration from 0.1 to 0.3 wt%, the slurries whose PAA had a medium or the lowest molecular weight maintained higher removal rates of Si_3N_4 film than did the slurry whose PAA had the highest molecular weight. In other words, with increasing PAA concentration and the addition of PAA having the same molecular weight, the removal rates of Si_3N_4 film for all slurries were markedly reduced, and they very quickly saturated at a higher molecular weight. By comparing Figures 3.26a and b, we can calculate the removal selectivity of the SiO_2-to-Si_3N_4 films. For the highest PAA molecular weight (MW = 90,000), the selectivity increased approximately from 10:1 to 13:1 with increasing PAA concentration. For the lowest molecular weight (MW = 30,000), however, the selectivity increased approximately from 4:1 to 51:1.

To clarify these results, the slurry samples were used in the STI planarization step for actual patterned wafers. Figure 3.27 shows the removed amounts of SiO_2 and Si_3N_4 films versus the pattern density of the patterned wafer for different PAA molecular weights and primary abrasive sizes. The SiO_2 film was fully overpolished with increasing pattern density, as shown in Figure 3.27a. The removed amount of Si_3N_4 film increased with increasing pattern density throughout the experimental range of PAA molecular weights on the concentration of 0.42 wt%, as shown in Figure 3.27b. In addition, as contrasted with the blanket wafer tests, with a higher PAA molecular weight and addition of the same PAA concentration, the removed amount of Si_3N_4 film in active regions was gradually reduced for all slurries. With a low PAA molecular weight, however, a smaller primary abrasive size maintained a higher removed amount of Si_3N_4 film at the same PAA molecular weight and concentration. Kang et al. (2004) reported that the abrasive size influences the effect of the PAA on the removal rate of a ceria slurry. They explained this result by using a model with the layer of PAA adsorbed or segregated on the film surface: larger abrasives are more likely to penetrate the viscous layer of adsorbed PAA, contact the hydrated surface, and form covalent bonds like Ce–O–Si on the film surface. According to this mechanism, the particle size determining the possibility of penetrating the viscous layer, contacting the hydrated film, and removing the film surface is one of the most important factors affecting the removal rate. As the particle size decreases, therefore, the removal rates also decrease. On the other hand, with many small particles remaining in the slurry suspension, whose surface areas are so large as to easily cause greater adsorption of

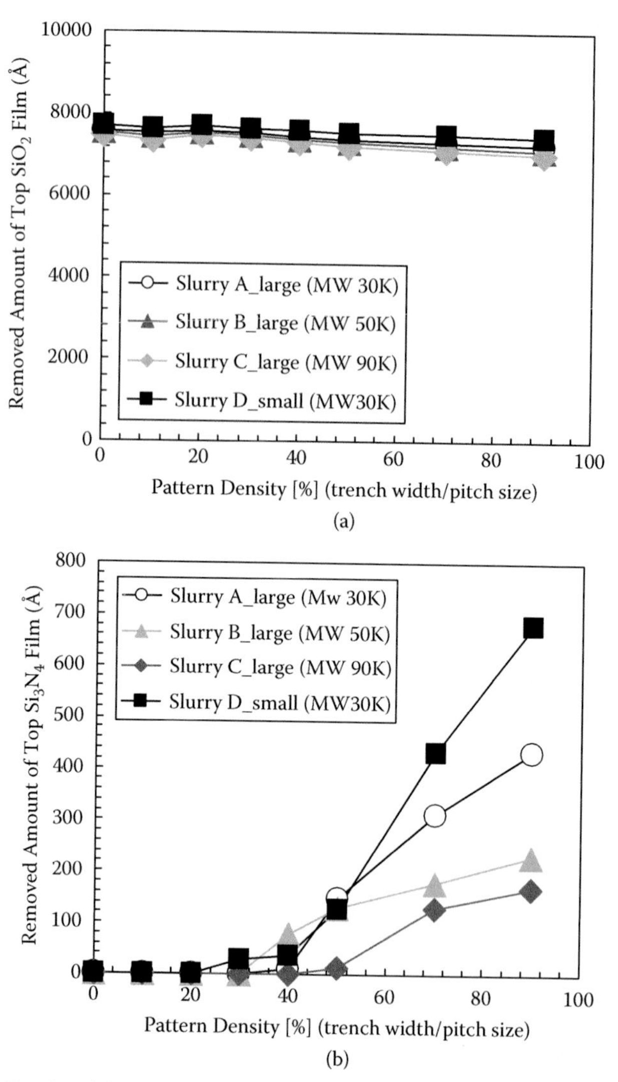

FIGURE 3.27 Results of the CMP tests of patterned wafers in terms of the surfactant molecular weight: (a) removed amount of SiO_2 film and (b) removed amount of Si_3N_4 film.

PAA molecules in the slurry, the Si_3N_4 film can easily be removed because of the PAA adsorbed insufficiently on the densely separated Si_3N_4 film surface on a patterned wafer.

Figure 3.28 shows cross-sectional SEM images of the 5-μm pitch size with the density fixed at 50%, illustrating the edges of active Si_3N_4 and trench SiO_2 layers before and after polishing. With a higher PAA molecular weight and the same PAA concentration, the removed amount of Si_3N_4 film for all three slurries was gradually reduced with the narrow pitch size of 5 μm. The Si_3N_4 film erosion was clearly less for the PAA with the highest molecular weight, as compared to that for the low molecular

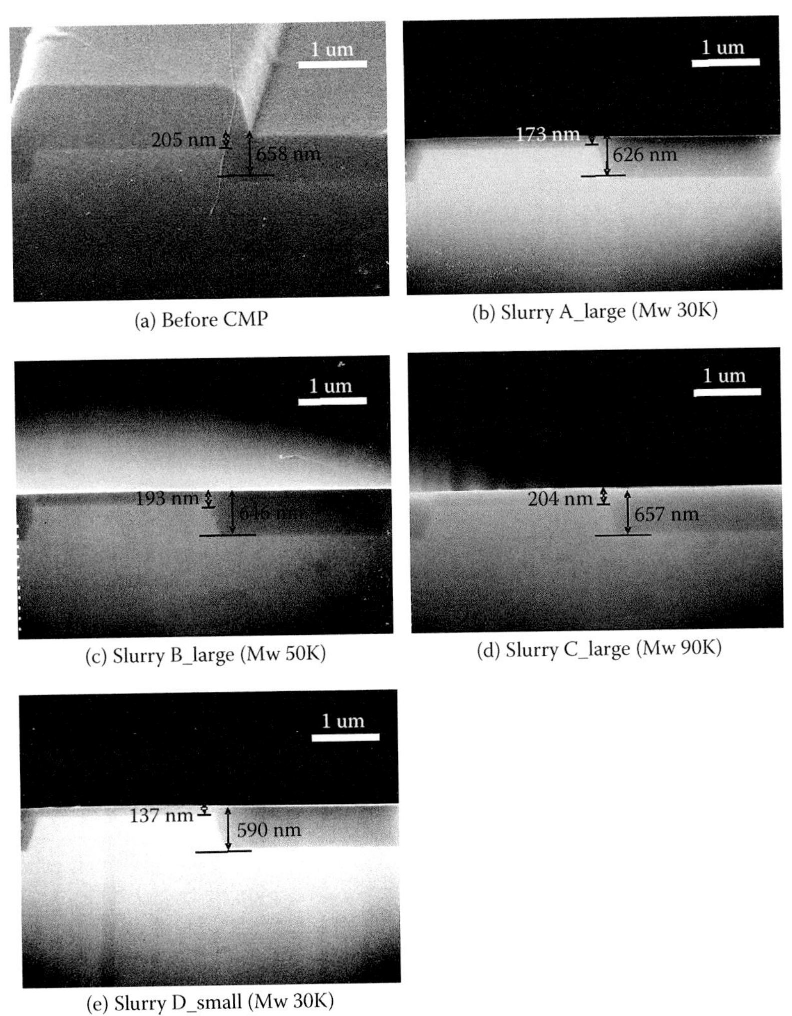

FIGURE 3.28 (See color insert) Pre- and post-CMP cross-sectional SEM micrographs: (a) pre-CMP, (b) slurry A_large (Mw 30K), (c) slurry B_large (Mw 50K), (c) slurry C_large (Mw 90K), and (d) slurry D_small (Mw 30K).

weight. Hence, at the same molecular weight, a smaller primary abrasive size maintained higher erosion of Si_3N_4 film than with a large primary size. The amount of PAA adsorption on the smaller particles was much higher than that on the larger particles because of their higher specific surface area, resulting in extra consumption of the PAA in the slurry solution. By comparing the images before and after CMP, we could calculate the amount of Si_3N_4 erosion. In this study, we also confirmed that the order of the measured Si_3N_4 film erosion (90K [large] < 50K [large] < 30K [large] < 30K [small]) did not change with respect to previous experimental results in this region with a low density of field Si_3N_4. These results are in good agreement with the Si_3N_4 film erosion shown in Figure 3.27b.

Figure 3.29a shows AFM line scan measurements indicating that a significant amount of SiO_2 local dishing occurred with overpolishing in a 500-μm-wide region. The dishing was reduced with increasing PAA molecular weight for the wide-field SiO_2 isolation region of 250 μm. Yu et al. (1992) explained the mechanism of the dishing effect. For a narrow field width, the pressure exerted on the field SiO_2 is significantly reduced when the interface between the SiO_2 and Si_3N_4 films is reached in the CMP process because the pressure applied by the pad is now concentrated on the Si_3N_4 layer as a result of its lower removal rate (about seven times lower than that of the field SiO_2). In the wide-field region, the reduction in the local pressure is far less significant because of the elasticity of the pad, resulting in continued polishing of the field SiO_2 after the film interface (i.e., between field SiO_2 and active Si_3N_4) is reached during CMP, so that the wider the field region, the smaller the reduction in the pressure acting on the field SiO_2, and the greater the degree of dishing. The dishing of the field SiO_2 was significantly lower because of the higher molecular weight of PAA in the ceria slurry, as shown in Figure 3.29a, which means that the PAA was more tightly adsorbed on the SiO_2 film because of the chain length and the chain bridging effect for the higher molecular weight than for the lower molecular weight. It was found that the surface roughness of the active region was much higher for the PAA with the highest molecular weight than for that with the low molecular weight, as shown in Figure 3.29b. The surface roughness of the active Si_3N_4 region became worse with a higher molecular weight and the same primary abrasive size and PAA concentration, while a low value for the surface roughness of the field SiO_2 region was maintained. We previously reported, according to AFM analysis, that the adsorption of anionic PAA is attributed to the formation of a PAA adsorption layer on the Si_3N_4 film, as a result of the electrostatic interaction between the PAA and the film surface.

The similar adsorption behavior of the PAA on the active Si_3N_4 films with a pattern density of 10% can be further characterized by the AFM images shown in Figure 3.30. The morphology and surface roughness dependencies after polishing of the active Si_3N_4 region on the different PAA molecular weights and primary abrasive sizes were observed. In the

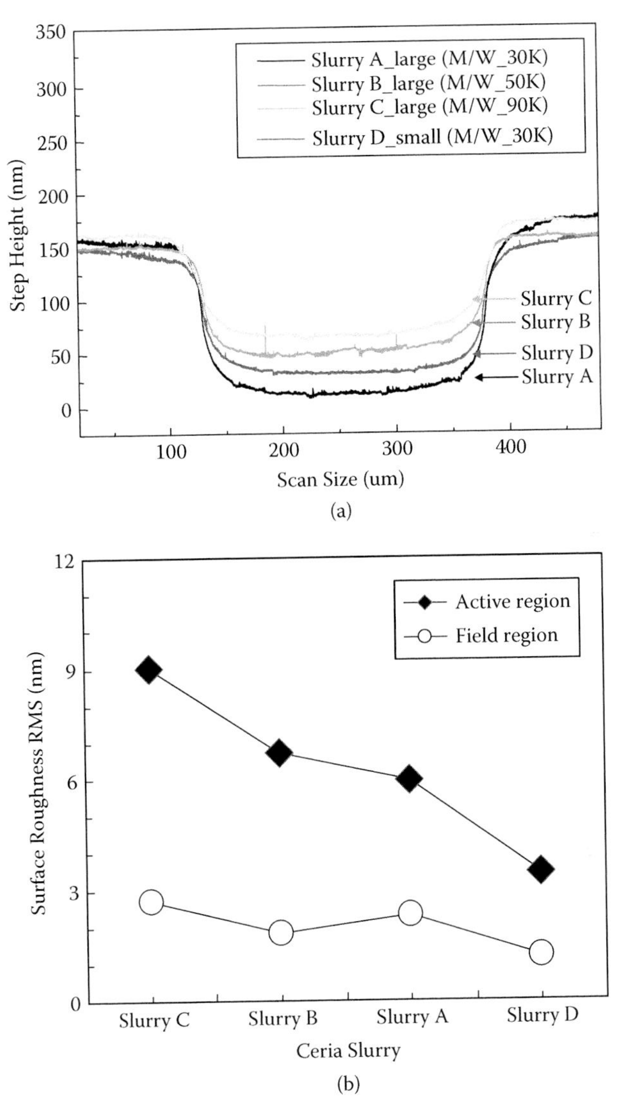

FIGURE 3.29 (a) Post-CMP surface line scans of the wide-field SiO_2 region; (b) post-CMP RMS surface roughness values.

case of the different primary sizes, there was no significant change in the film surface between the Si_3N_4 active surface and the adsorbed PAA after polishing. On the other hand, it was found that the surface roughness of the post-CMP Si_3N_4 film for the PAA with the highest molecular weight was much higher (0.355 nm) than that for the lowest molecular weight (0.280 nm), as illustrated in Figures 3.30a and 3.30d. This is attributed to the formation of the PAA adsorption layer on the Si_3N_4 film because of

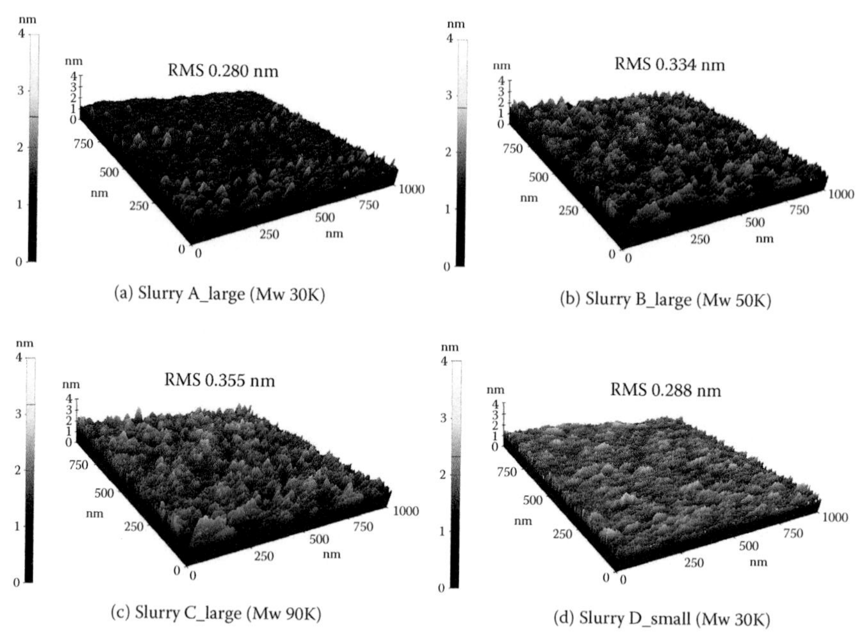

FIGURE 3.30 (See color insert) Post-CMP three-dimensional AFM micrographs of patterned wafers: (a) slurry A_large (Mw 30K), (b) slurry B_large (Mw 50K), (c) slurry C_large (Mw 90K), and (d) slurry D_small (Mw 30K).

the electrostatic interaction between the PAA and the film surface. These results are in good agreement with the AFM line scan measurements shown in Figure 3.29.

The adsorption behavior of the PAA on the ceria particles was mainly caused by the different surface charges between the PAA and the ceria surface. The PAA is an anionic polyelectrolyte with an acidic carboxyl group, which leads to the ionization of the PAA molecules in the neutral pH region at which the ceria slurry for STI CMP is usually manufactured. Meanwhile, the net surface charge of the ceria particle is near zero in this pH region because the pH_{iep} of ceria is approximately 6 to 7. Thus, the partially or fully ionized polyelectrolyte (PAA) is adsorbed on the ceria surface by electrostatic interactions. The electrostatic attractive force between adsorbed PAA molecules on the water-particle interface and the Si_3N_4 film surface can be classified as mainly resulting from the electrostatic interaction of the electric double layer surrounding the particles and the steric hindrance effect of the adsorbed PAA molecules on the Si_3N_4 film. Since the interaction of the electric double layer may have increased in proportion to the surface potential of absorbed PAA molecules with the same counterion content, the change in the zeta potential is important for the dependence of the electric double layer on the suspension properties. Moreover, the oxide film and the surface-modified ceria particles are negatively charged above pH 3, while the Si_3N_4 film is positively charged below pH 7. The attraction or repulsion between the abrasive particles and films (SiO_2 and Si_3N_4) results from the different electrostatic potentials exhibited in certain pH regions. Hence, during the blending of the slurry and additive solution, PAA that is used to form a passivation layer on the Si_3N_4 film can be additionally adsorbed on the surface of the ceria particles, which are basically covered by the same organic additive acting as the dispersant. This phenomenon could be explained as follows: the repulsive interaction between adjacent carboxyl sites is generated through the addition of more polymers, which then resulted in the conformational change of the adsorbed polymer. In addition, the carboxylic acid group appear to be necessary to suppress the Si_3N_4 removal rate during CMP process through hydrogen bonding between Si_3N_4 film and carboxylic group in amino-acid-based ceria slurry. The electrostatic interactions between the abrasive particles in each slurry and the film surfaces, however, may not fully explain the suppressed removal rate of Si_3N_4 film and the removal selectivity of SiO_2-to-Si_3N_4 films with different PAA molecular weights. Hence, it is necessary to consider other factors that influence the abrasive movement in a slurry, from the point of view of rheological behavior. These factors depend on the passivation layer of PAA at the interface between the film surface and the ceria slurry suspension.

Figure 3.31 shows the rheological behaviors of various ceria slurries, with a fixed PAA pH of 7.0, as a function of the PAA concentration and molecular weight with different primary abrasive sizes. As shown in Figure 3.31a, for the PAA with the higher molecular weights, the slurry viscosity increased markedly with the PAA concentration, but it barely increased with the concentration in the case of the low molecular weight, regardless of the abrasive size. The primary abrasive size made no difference in the average slurry viscosity for the two different primary sizes. The measured effects of the PAA concentration and molecular weight on the slurry viscosity are in good agreement with results previously reported. For the same weight concentration of ceria abrasives, the number of molecules in a ceria suspension with a higher molecular weight PAA will be lower than that with a lower molecular weight.

According to the Mark–Houwink–Sakurada equation, the relationship between the viscosity, molecular weight, and organic polymer type can be formulated as

$$[\eta i] = Km\, M_w{}^a \tag{3.4}$$

where ηi is the intrinsic viscosity, a and Km are constants for a specific polymer solvent system, and M_w is the average molecular weight of the polymer. The constant Km depends on primary molecular features, such as the persistence length, while a depends on short-range interactions and their implied effect on the molecular weight. For each slurry with a range

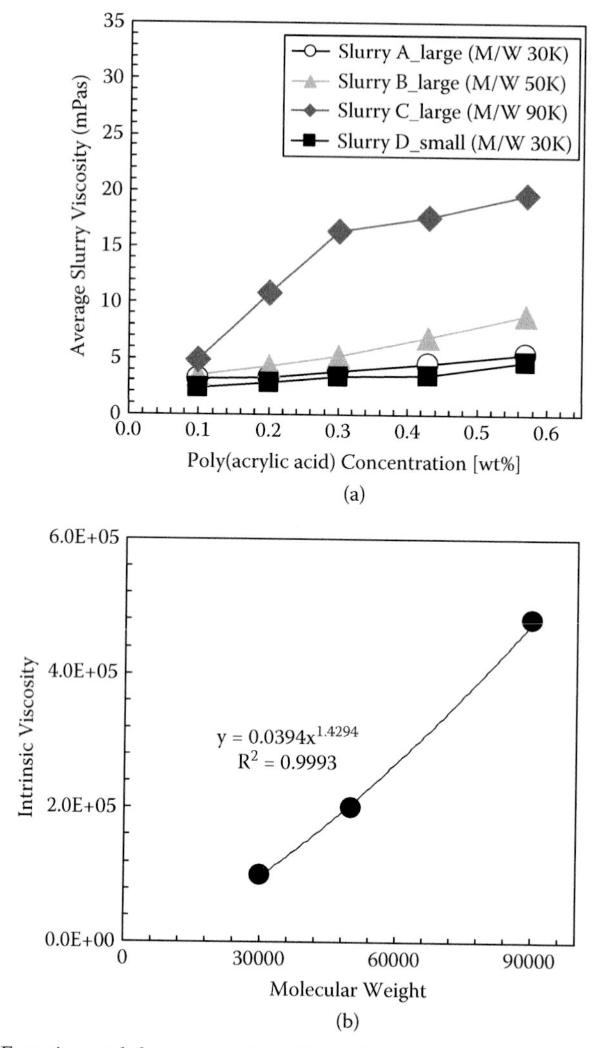

FIGURE 3.31 Experimental slurry viscosity with surfactants of different molecular weights: (a) average slurry viscosity as a function of the surfactant concentration and molecular weight, and (b) intrinsic viscosity calculated from eqs. 1 and 2.

of PAA molecular weights, the relation between the intrinsic viscosity and molecular weight is one of its most important properties. This relation can be represented by the following equation:

$$ln\ \eta i = ln\ Km + a\ ln\ M_w \tag{3.5}$$

Figure 3.31b shows the calculated parameters *Km* and *a* from a plot of *ln* *ηi* versus *ln Mw*. For the constants in Equation 3.5, we chose the values for

PAA in a ceria slurry solution, enabling us to evaluate the average molecular weight of PAA in this solution. Here, the amount of PAA adsorption (or segregation on the surface) depends on the bulk concentration of the PAA and the electrostatic interaction between the PAA and the film surface. Moreover, because a PAA with a higher molecular weight adsorbs more densely, the intrinsic viscosity (ηi in Equation 3.4), which describes the particle behavior near the film surface, should increase and hinder the movement of particles. As a result, the removal rates of both the SiO_2 and the Si_3N_4 films were reduced as the molecular weight and the concentration of the PAA increased in the blanket wafer tests. Moreover, the removal rates of the SiO_2 and Si_3N_4 films can become important, depending on the passivation layer of PAA existing at the interface. Thus, the addition of a PAA with a lower molecular weight appears to passivate the electrostatic interactions, thereby resulting in weaker adhesion of the adsorbed PAA layer through polymer chain bridging and branching, and possibly resulting in desorption of this layer above a certain applied load during the CMP process. As the PAA chain length increases, however, the lateral interaction among the hydrocarbon chains becomes more pronounced, resulting in the formation of a more effectively passivated layer of PAA. Consequently, with increasing PAA concentration and addition at a higher molecular weight, the Si_3N_4 removal rates for all slurries markedly reduced in the blanket wafer tests. Although in the case of a higher PAA molecular weight, the removal rate and erosion of Si_3N_4 film could be reduced, and for a patterned wafer, the removal rate of the field SiO_2 film was also reduced, while the surface roughness of the Si_3N_4 film in the active region was increased.

For the PAA with the highest molecular weight in our experiments using blanket wafer, with different primary abrasive sizes, the removal rates of the SiO_2 and Si_3N_4 films were markedly reduced as the PAA concentration increased. Hence, with increasing PAA concentration, a higher primary abrasive size maintained a higher removal rate of SiO_2 at the same PAA molecular weight and concentration. For the case of patterned wafers, with a higher PAA molecular weight, the erosion of Si_3N_4 film could be reduced, but our pattern wafer tests showed that the removal amount was reduced and the surface roughness of the Si_3N_4 film became worse. These results can be qualitatively explained from the layer of PAA adsorbed on the film surface in terms of electrostatic interaction and rheological behavior, including the molecular weights, concentrations of PAA, and different primary abrasive sizes in the ceria slurry.

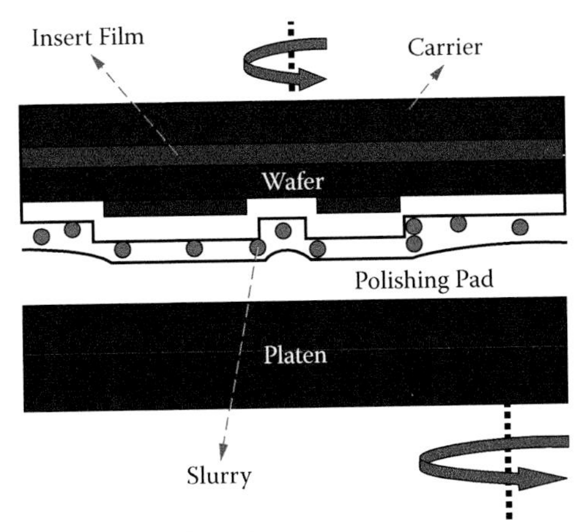

FIGURE 1.5 CMP process of manufacturing.

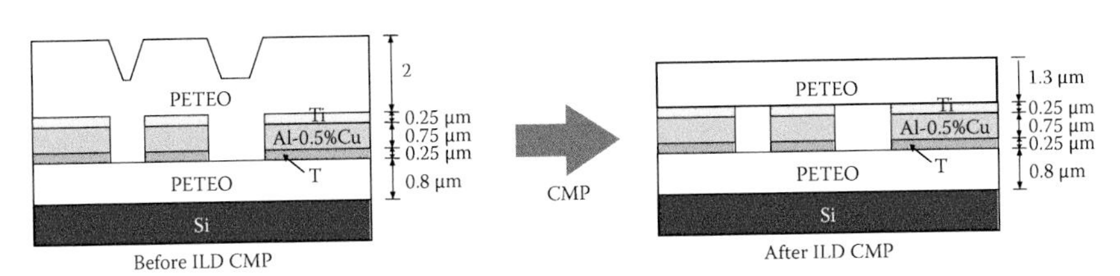

FIGURE 2.1 Schematic of ILD CMP process.

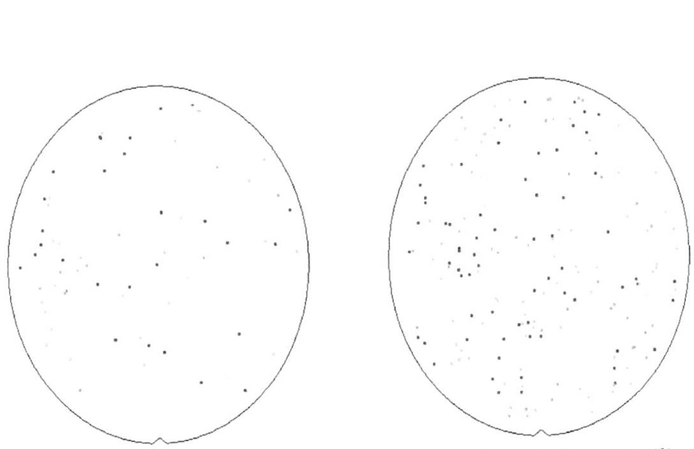

FIGURE 2.7 Analysis of remaining silica particles (particle size > 0.189μm) on silicon wafers after post CMP cleaning: (left) modified slurry, (right) nonmodified slurry.

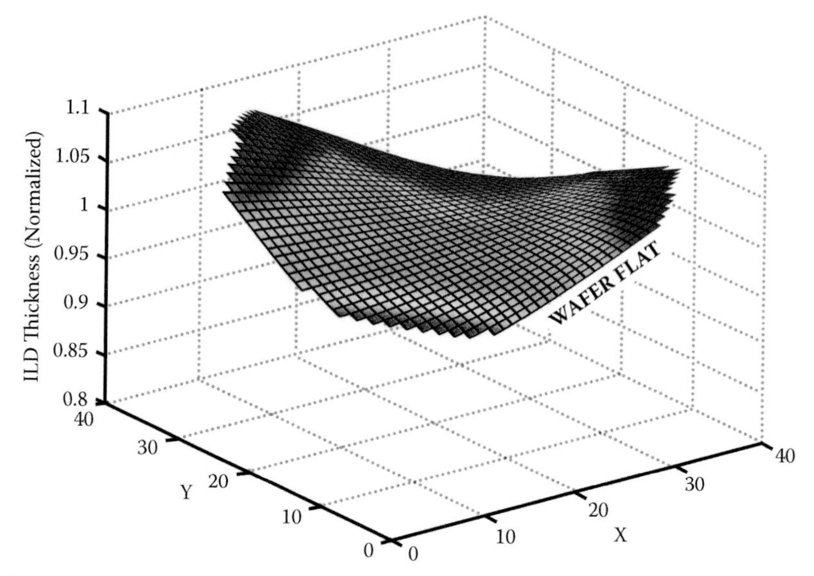

FIGURE 2.16 Wafer level variation for tool A.

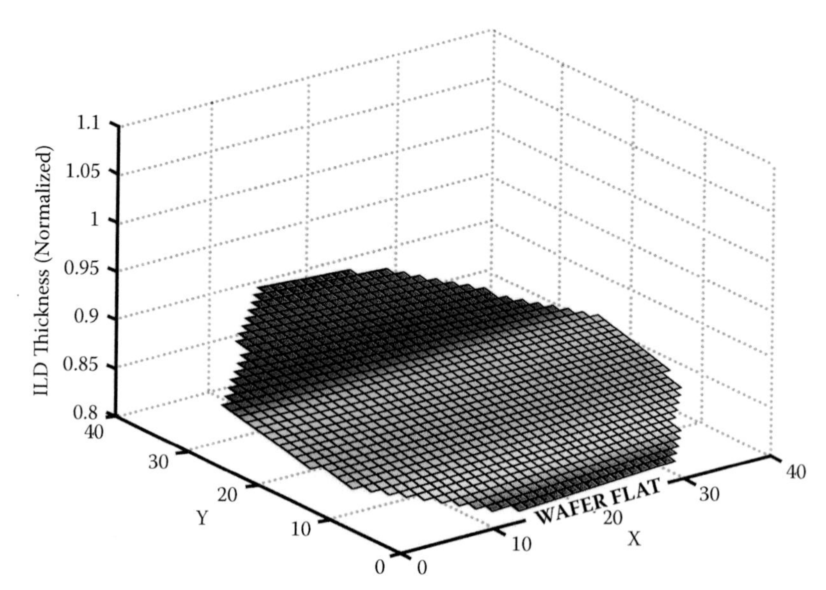

FIGURE 2.17 Wafer level variation for tool B.

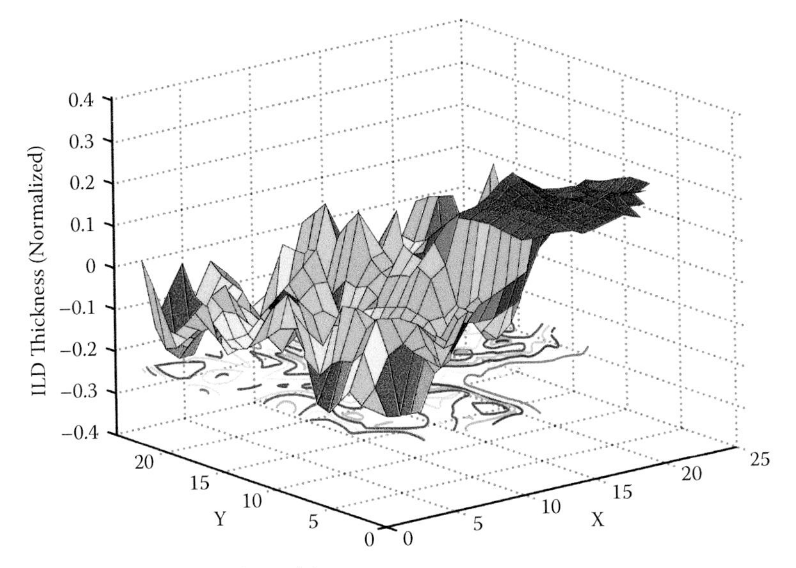

FIGURE 2.18 Die variation for tool A.

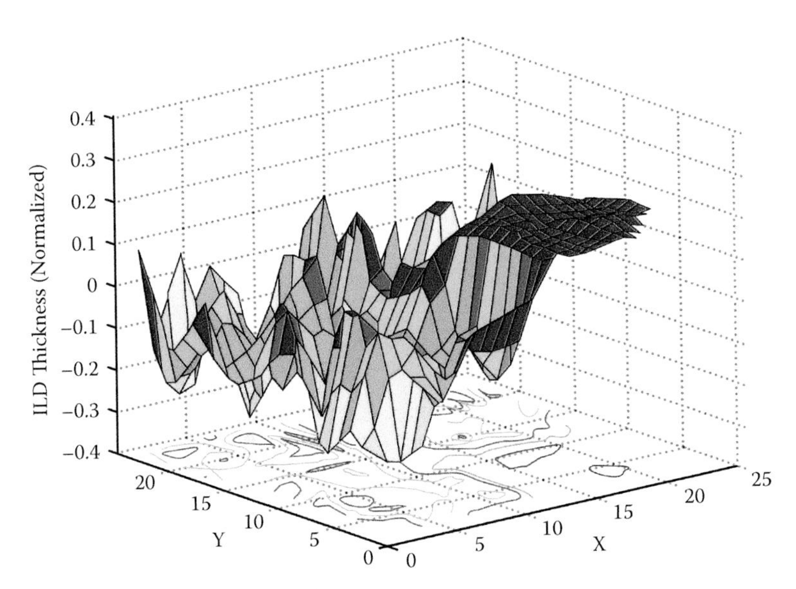

FIGURE 2.19 Die variation for tool B.

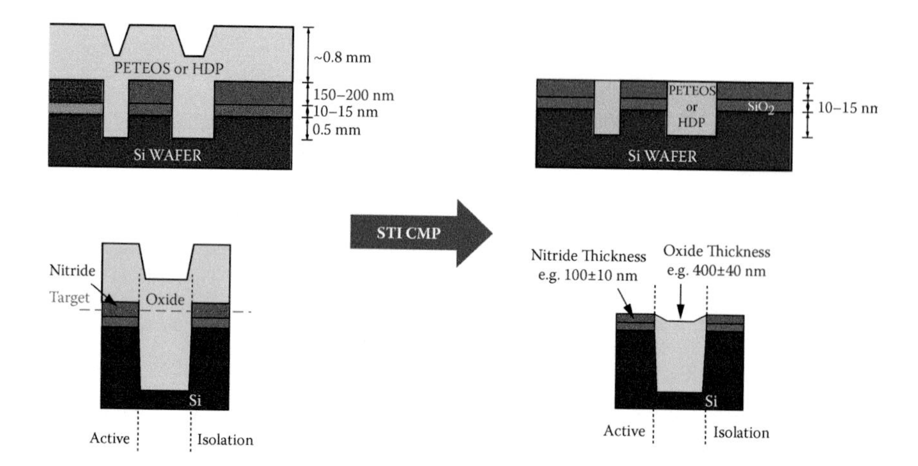

FIGURE 3.1 STI CMP process.

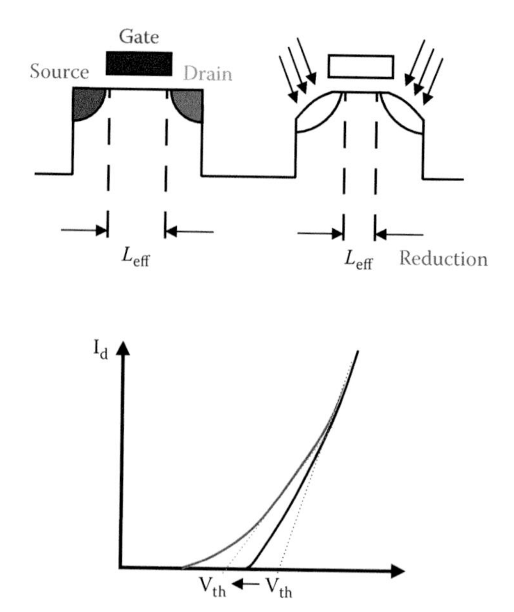

FIGURE 3.2 The effect of overpolishing on the characteristics of the device: (top) the decrease of gate length by overpolishing; (bottom) the shift of threshold voltage in the device.

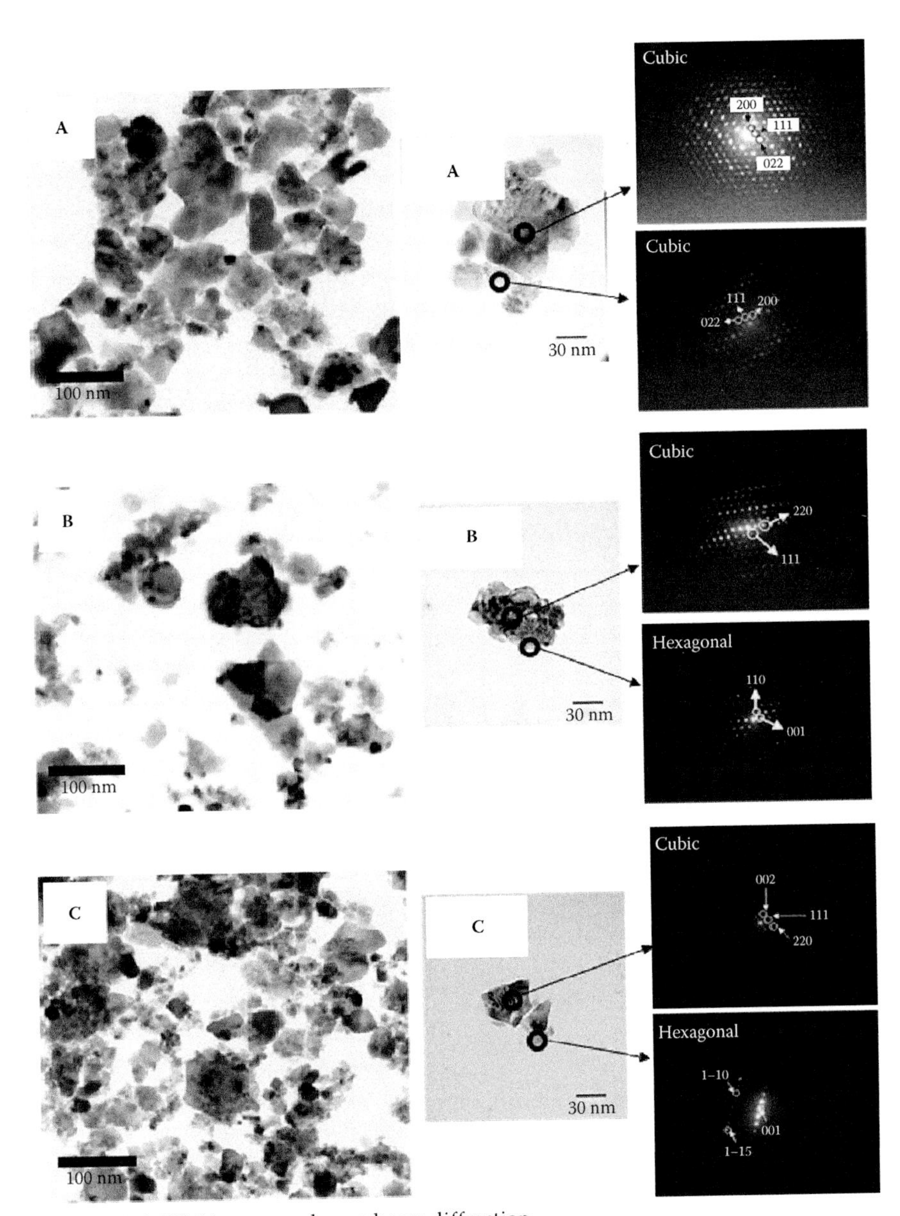

FIGURE 3.6 TEM images and nano beam diffraction.

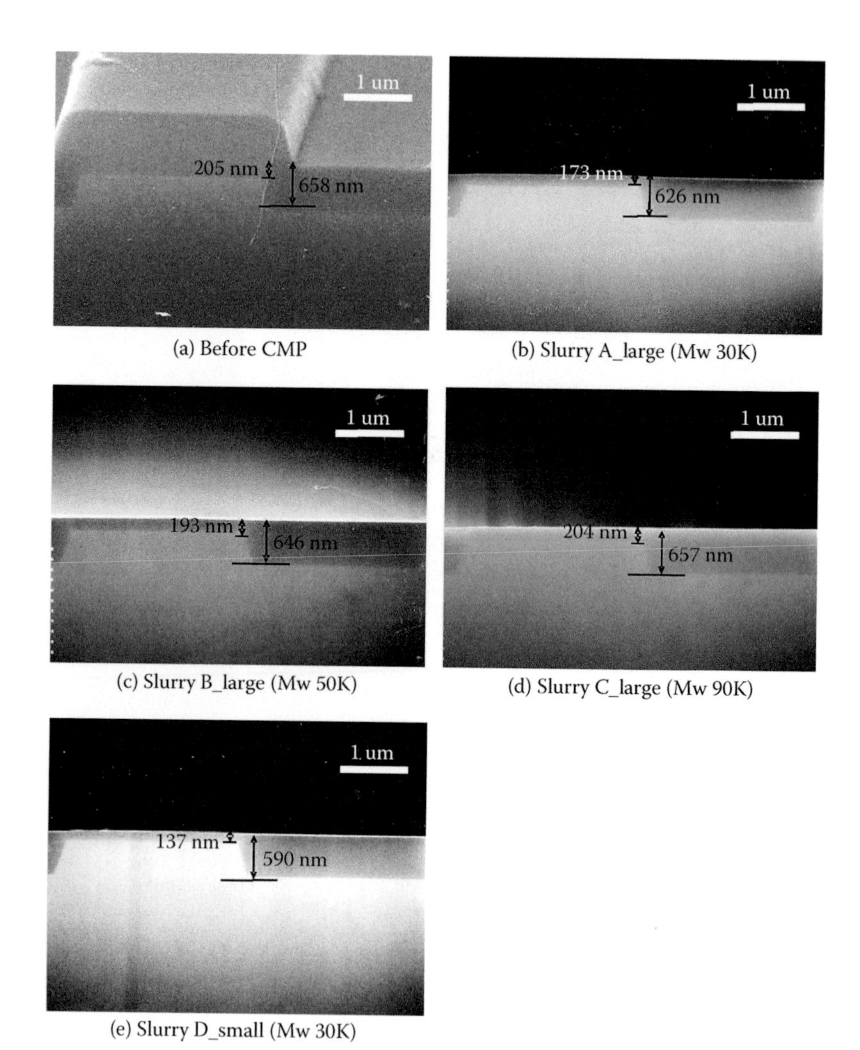

FIGURE 3.28 Pre- and post-CMP cross-sectional SEM micrographs: (a) pre-CMP, (b) slurry A_large (Mw 30K), (c) slurry B_large (Mw 50K), (c) slurry C_large (Mw 90K), and (d) slurry D_small (Mw 30K).

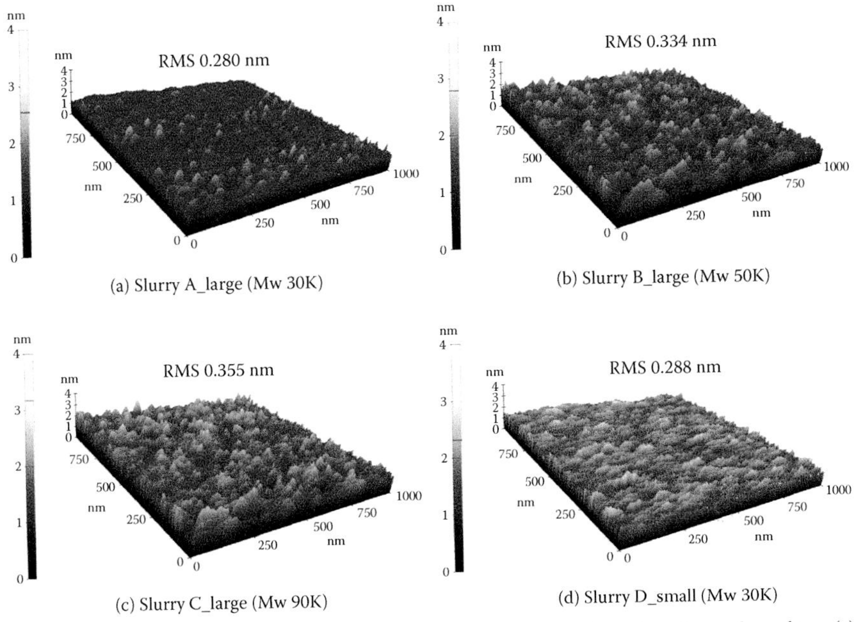

FIGURE 3.30 Post-CMP three-dimensional AFM micrographs of patterned wafers: (a) slurry A_large (Mw 30K), (b) slurry B_large (Mw 50K), (c) slurry C_large (Mw 90K), and (d) slurry D_small (Mw 30K).

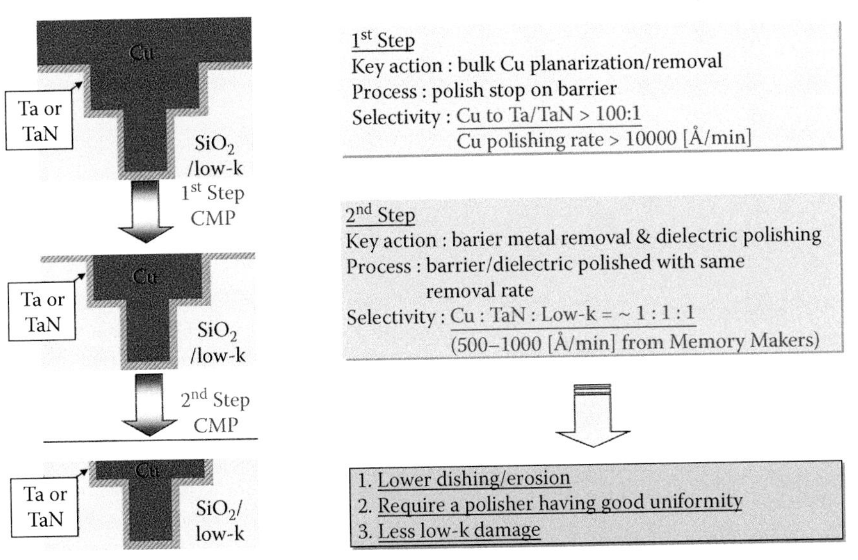

FIGURE 4.2 Two-step process of Cu CMP.

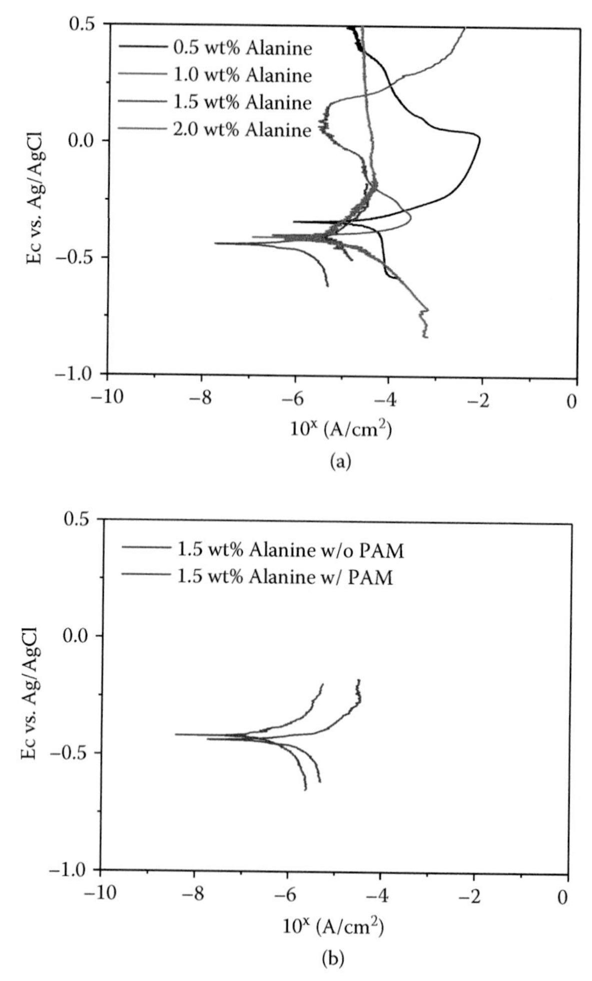

FIGURE 4.10 Potentiodynamic polarization: (a) various alanine concentration, (b) 1.5 wt% alanine with and without PAM.

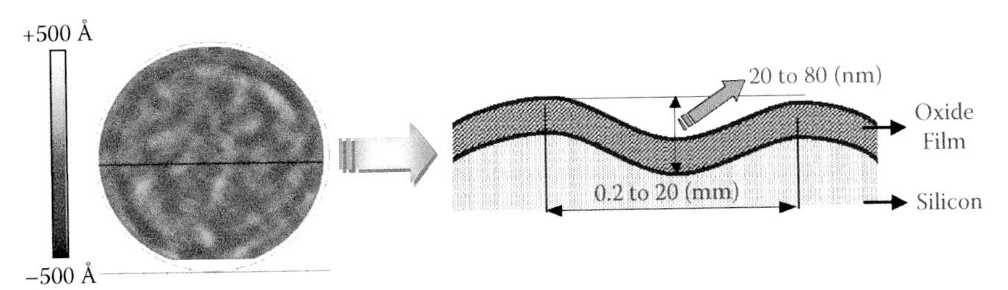

FIGURE 5.1 Illustration of wafer nanotopography.

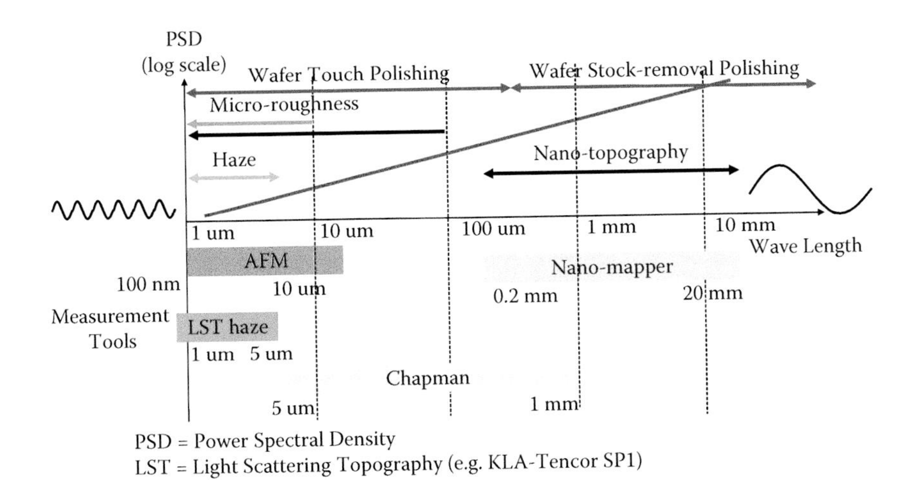

FIGURE 5.3 Topology map of wafer surface.

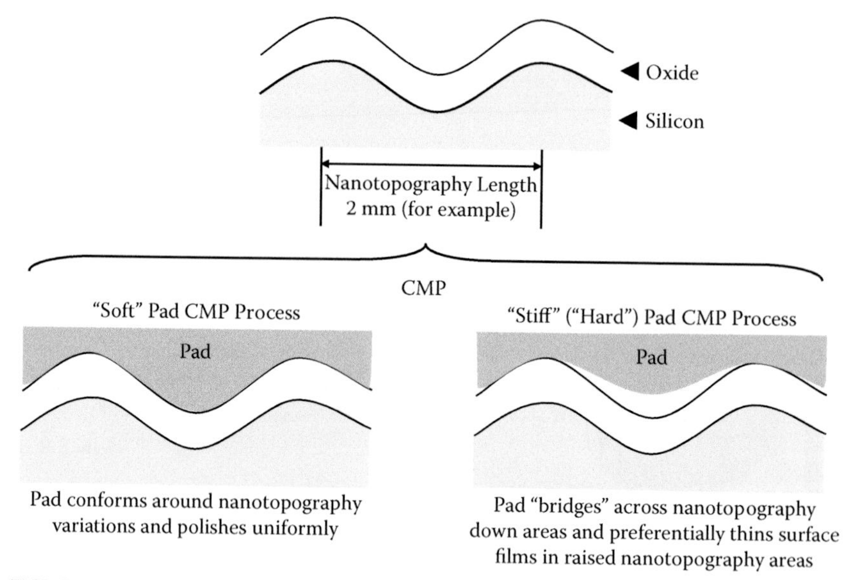

FIGURE 5.4 Basic concepts of soft and hard polishing pads.

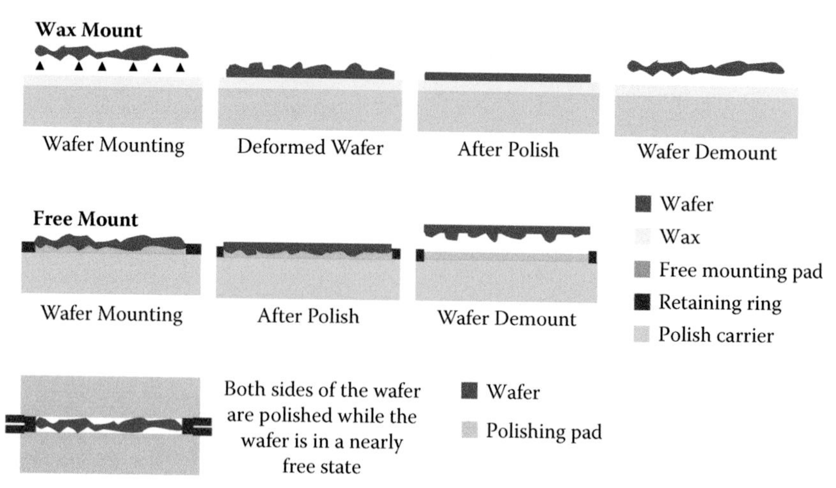

FIGURE 5.6 A comparison of SSP and DSP mounting techniques and how these affect nanotopography and flatness.

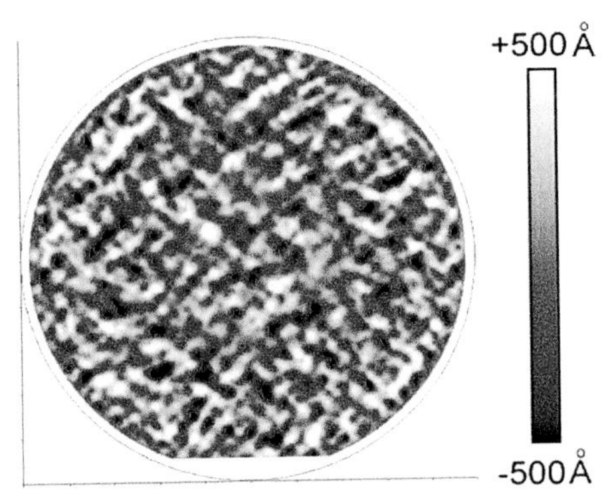

FIGURE 5.7 Nanotopography map for a wafer.

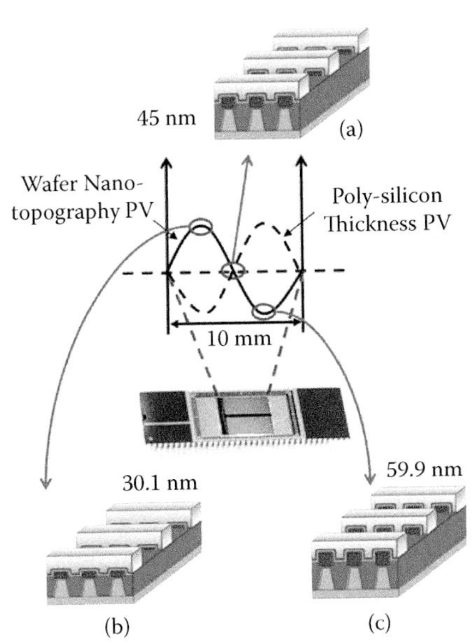

FIGURE 5.21 Schematic structures of a 63-nm NAND-flash memory-cell with different floating-gate heights, induced by the wafer nanotopography of 10-mm-diameter scanning: (a) 45-nm height (free of wafer nanotopography influence); (b) 30.1-nm height (at the top of wafer nanotopography influence); (c) 59.9-nm height (at the bottom of nanotopography influence).

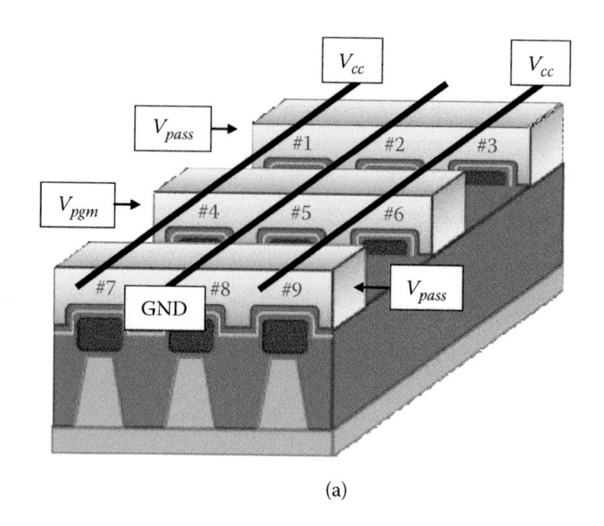

(a)

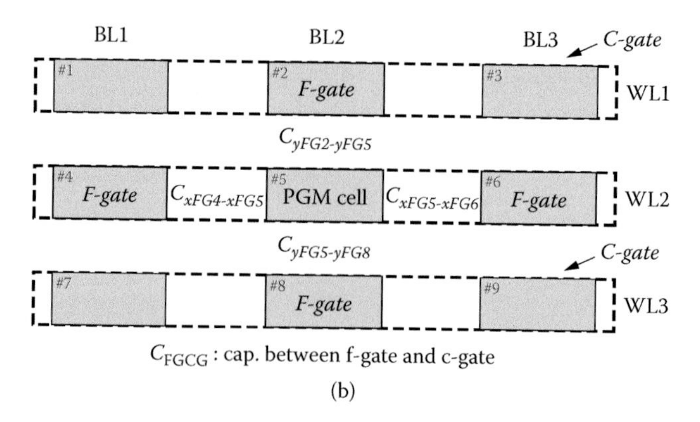

(b)

FIGURE 5.22 Programming operation of a NAND-flash memory-cell, where the programming cell is No. 5: (a) voltage bias conditions for memory cells and (b) parasitic capacitances during No. 5 cell programming.

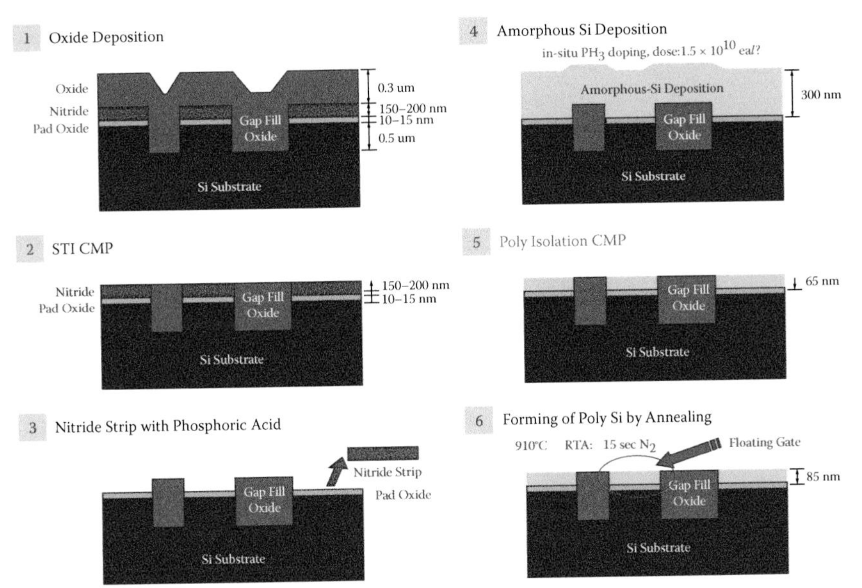

FIGURE 6.9 Schematic process flow of the poly isolation CMP process.

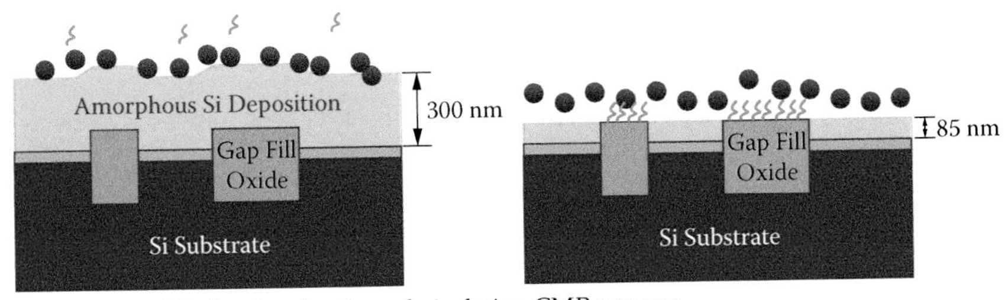

FIGURE 6.10 Mechanism for the poly isolation CMP process.

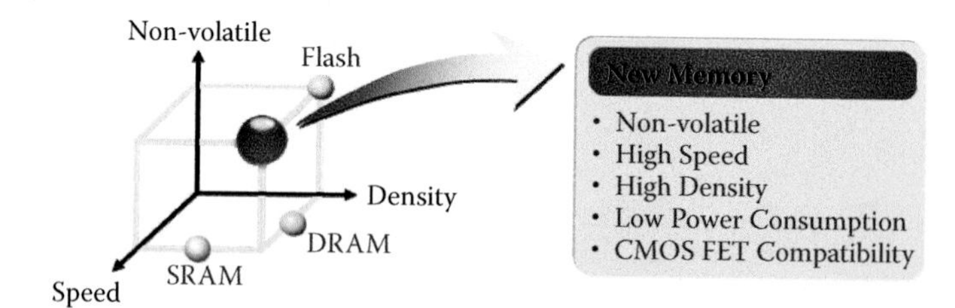

FIGURE 6.16 Comparison of new memory and conventional memory.

4

Copper CMP

4.1 Introduction

As design technology and manufacturing process technology of ultra large scale integration (ULSI) technology are improved, integration doubles every three years. As the design rule of semiconductor devices decreases below 100 nm, device is becoming high integration and multi-layer metallization of circuits. Especially as the feature size of the transistor is decreased below 130 nm, the device operation should be improved. However, parasitic capacitance that is caused by the reduction of the gap between metal lines increases, which leads to the decline of device operation. Therefore, copper (Cu) wire is used instead of aluminum (Al) wire to solve this problem.

In the case of microprocessor, if tungsten (W) plug and aluminum metal line is used, metal wiring of 10×12 layers is need. However, the application of copper wiring enables 6×8 metal layers. This can prevent the operation speed of device by the use of low-k dielectric to insulator of metal wiring instead of silicon oxide (SiO_2), because copper has low resistivity and superior electromigration. These can be explained by the delay time equation:

$$\text{delay time} \propto RC \propto \rho o \ell k \varepsilon o \tag{4.1}$$

where, R is resistance of metal wiring, and C is parasitic capacitance between metal wiring, ρo is resistivity of metal wiring, ℓ is the length of metal wiring, k is the permittivity of insulator between metal wiring, and εo is vacuum permittivity.

If the design rules of a semiconductor device gets into 0.1 μm low, element delay time is expected to be decreased with decreasing of gate. However, parasitic capacitance by the reduction of space length between Al metal wiring increases rapidly and delay time of the device increases rapidly. To solve this problem, the metal line/insulator substitutes Cu/low-k for Al/SiO_2, which results in a minimization of RC delay-time (Figure 4.1).

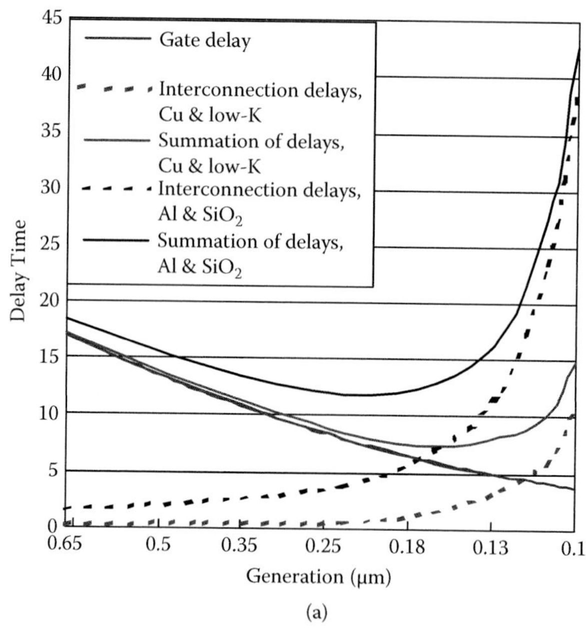

(a)

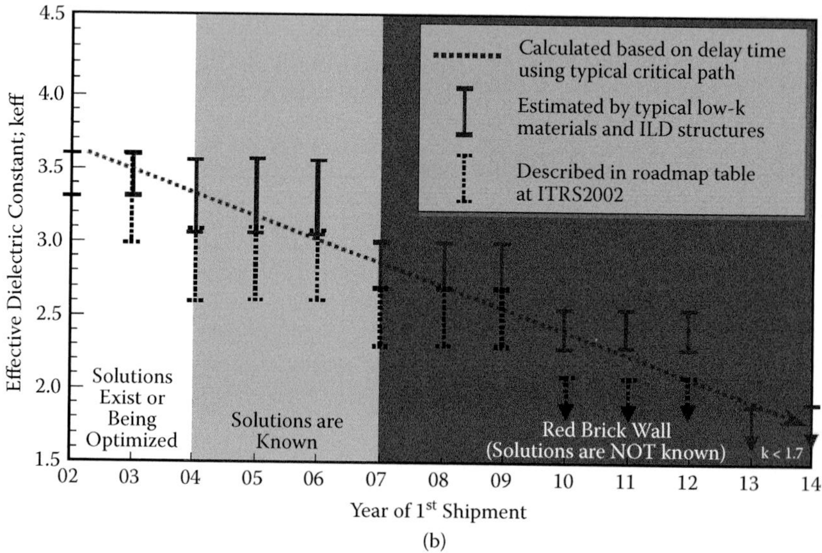

(b)

FIGURE 4.1 (a) Required delay time and (b) permittivity by design rule.

Barrier metal such as Ta or TaN prevents the diffusion of Cu because Cu's solubility and diffusivity are high when wire that uses Cu is required. And also because Cu and low-k material are very soft, there is difficult point that CMP process must perform under low shear stress and pressure.

Because Cu-to-TaN removal selectivity is not high, dishing or erosion happens. As these phenomenon displays become serious, the tendency aspect ratio does increase. Therefore, Cu-to-TaN removal selectivity is required more than a given level in CMP. Because more than 20:1 of removal selectivity is required, dishing does not happen. Also, to prevent erosion, 1:1:1 of Cu:Ta:low-k removal selectivity is required as shown in Figure 4.2 and Figure 4.3. Dishing or erosion, which is caused after Cu CMP, can have an adverse effect on device operation, such as an increase of resistance.

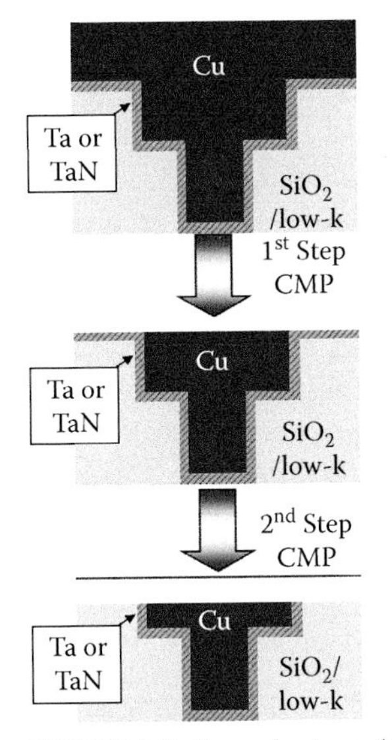

1st Step
Key action : bulk Cu planarization/removal
Process : polish stop on barrier
Selectivity : Cu to Ta/TaN > 100:1
Cu polishing rate > 10000 [Å/min]

2nd Step
Key action : barrier metal removal & dielectric polishing
Process : barrier/dielectric polished with same removal rate
Selectivity : Cu : TaN : Low-k = ~ 1 : 1 : 1
(500–1000 [Å/min] from Memory Makers)

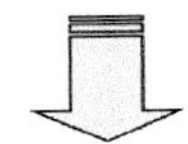

1. Lower dishing/erosion
2. Require a polisher having good uniformity
3. Less low-k damage

FIGURE 4.2 (See color insert) Two-step process of Cu CMP.

YEAR		09	10	11
Memory Technology Node (nm)		45	40	36
Cu Metallization (Cu-CMP)				
1st Step Slurry	Cu Polishing Rate (Å/min)	10000±500	12000±400	15000±400
	Barrier Polishing Rate (Å/min)	70±10	60±5	50±5
	Selectivity (Cu vs. Barrier)	>100	>120	>160
2nd Step Slurry	Polishing Rate (Å/min)	500:500:500	250:500:500	250:500:500
	Selectivity (Cu : Barrier : Low-k)	1:1:1	0.5:1.1	0.5:1:1
Remaining Particle	Size (um)	0.1		
	(cm^{-2})	≤0.17	≤0.17	≤0.17
	(#/wafer)	≤0.16	≤120	≤115
Scratch (#/cm^2)		1um		
		0.010	0.006	0.003
Cu Dishing (Å)		400	300	200
Cu Erosion (Å)		300	200	150

FIGURE 4.3 Cu CMP roadmap.

4.2 High Selectivity for Copper CMP

Cu CMP slurries commonly use submicron-sized colloidal silica abrasive particles dispersed in aqueous solutions that contain an oxidizer, as well as an complexing agent and corrosion inhibiting agents and other chemicals. Most of the slurries described in the article by several researchers use H_2O_2 as the oxidizer and benzotriazole as the inhibitor, with various complexing agents such as organic polymer, alkaline agent, and organic amine in slurry.

The pH value of the polishing slurry is one of the most important parameters influencing the polishing rate, surface roughness, and other performance characteristics of the Cu CMP process. In this section, the slurry's pH and conductivity were adjusted to the range of pH 10 to 11 and conductivity of 8 to 10 (mS/cm) by adding an alkaline agent, including NH_4OH and HCl solution.

Figures 4.4 to 4.7 show the results obtained from an experiment conducted by varying the concentrations of complexing agent (alanine) and selectivity control agent (PAM) in aqueous slurry. Figure 4.4 shows the removal rate of Cu and TaN films versus the alanine concentration. The removal rate of Cu film increased with alanine concentrations. In addition, the removal rate of TaN film strongly suppressed and slightly increased with increasing alanine concentrations in aqueous suspension. As with the removal rate of Cu film, the removal rate of TaN film drastically decreased and was essentially saturated with a concentration of alanine beyond 0.5 wt%.

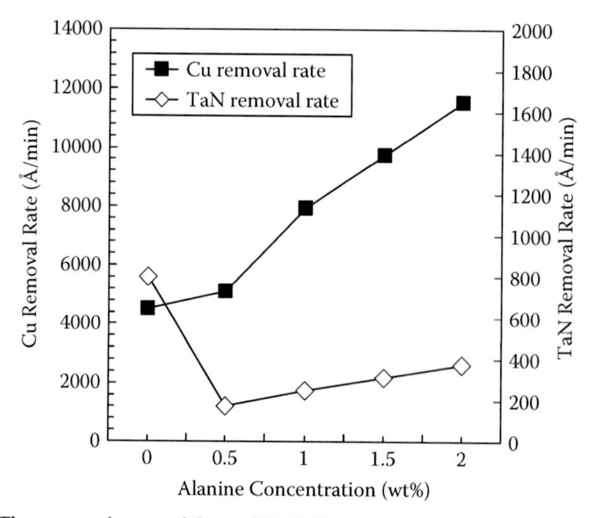

FIGURE 4.4 The removal rates of Cu and TaN films versus alanine concentration in slurry.

Alanine could exist in aqueous solution in three different forms, namely, CH3 CH(NH3+)COOH (cation), CH3 CH(NH3+)COO– (zwitterions), and CH3 CH(NH2)COO– (anion). These species are denoted as H2L+, HL, and L–, respectively, for brevity. The equilibrium between these may be depicted, as Babu et al. (2005) previously reported, the dissolution and removal rate probability of the complexing agent, including phthalic acid, citric acid, glycine, oxalic acid, and carboxyl and/or amine functional group, which interact on the Cu film surface should strongly influence the removal rate.

$$\underset{\substack{(H2L^+)\\ \text{cation}}}{CH3\ CH(NH3^+)\ COOH} \xleftrightarrow{pKa1 = 2.35} \underset{\substack{(HL)\\ \text{zwitterion}}}{CH3\ CH(NH3^+)\ COO^-}$$

$$\xleftrightarrow{pKa2 = 9.87} \underset{\substack{(L^-)\\ \text{anion}}}{CH3\ CH(NH2)\ COO^-} \tag{4.2}$$

Babu et al. explained that a complexing agent, such as amino group in glycine and hydrogen peroxide system, is protonated at pH <4.0, and thus may not effectively form chelates with positively charged metal ions; thus, the dissolution must be due to the carboxyl group. On the other hand, at pH >4.0 the amino group can chelate Cu2+ ion and cause the dissolution of the metal up to pH 10. However, alanine and H_2O_2–containing colloidal silica slurry exhibited an enhanced removal rate of Cu film at alkaline pH region. We thought that the alanine could be a very effective complexing agent with an increased removal rate of Cu film through a high dissolution rate of Cu2+ ion in alkaline pH region.

The suppression of the removal rate of TaN film could not be fully explained through the electrochemical phenomena by chemical reaction between complexing agent and the TaN film surface. We thought that the TaN film loss and the Cu-to-TaN removal selectivity are directly related to the electrostatic interaction and electrokinetic behavior due to chemical adsorption and steric hindrance of adsorbed organic chemical.

Figure 4.5 shows the electrokinetic behaviors of Cu film, TaN film, and colloidal silica slurries with alanine addition as a function of pH. The absolute surface zeta potential of the Cu film was slightly negatively charged above pH 5. The TaN film also exhibited a slightly negative charge at a pH above pH 5.3. Colloidal silica slurry with alanine exhibited a pH_{iep} at pH 4.0.

The surface potentials of the colloidal silica abrasive particles in the aqueous suspension with alanine were strongly negatively charged above pH 4.0, while the TaN film's surface potential was weakly negatively charged. The attraction behavior between the abrasive particles and the TaN film results from the different electrostatic potentials exhibited in certain pH regions. Therefore, we suggest that the selective adsorption of alanine added slurry on the abrasive particles and the TaN film surfaces correspond to the differing zeta potential charge. The removal rate of TaN film is drastically supressed, and increased slightly by this difference.

By comparing the removal rate of Cu and TaN films, we can calculate the removal selectivity of Cu-to-TaN films (Figure 4.6). By increasing

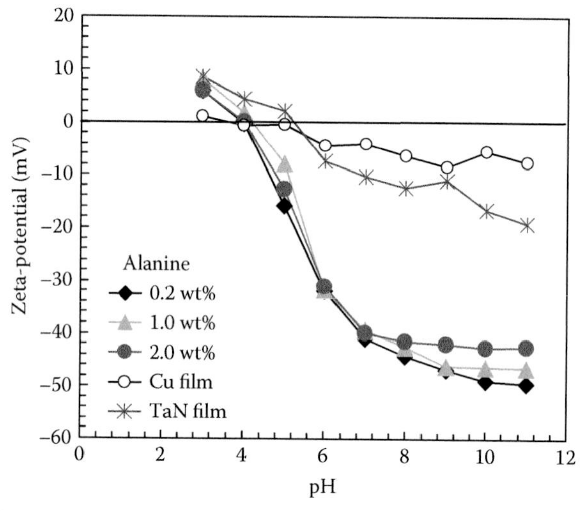

FIGURE 4.5 Zeta potential of Cu, TaN films, and colloidal silica slurry with alanine as a function of pH.

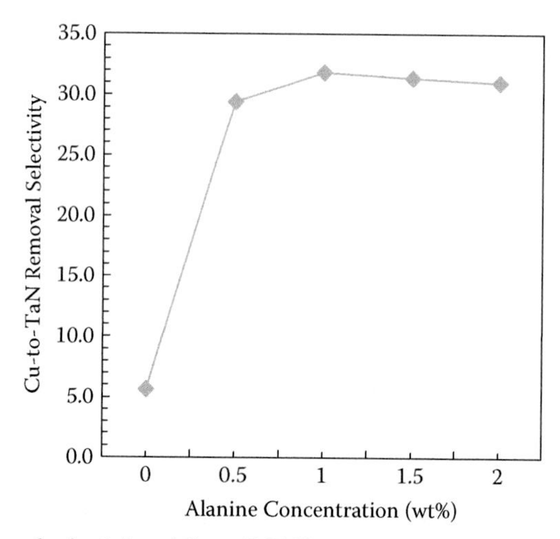

FIGURE 4.6 Removal selectivity of Cu-to-TaN films versus alanine concentration in slurry.

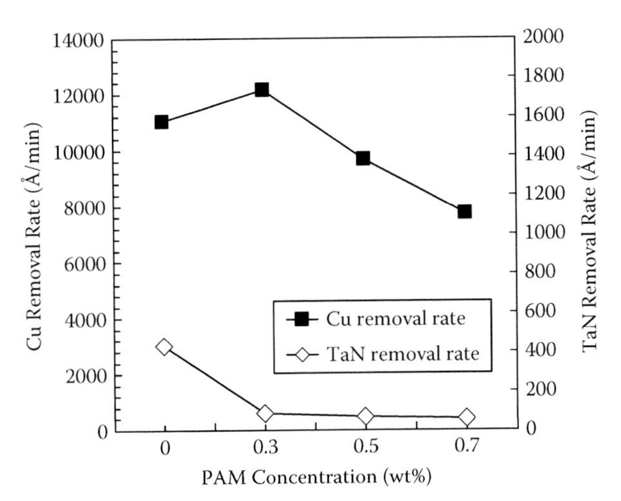

FIGURE 4.7 The removal rates of Cu and TaN films versus PAM concentration in slurry.

the alanine concentration, the removal selectivity drastically increased and essentially saturated from 5:1 to 32:1 with increasing alanine concentration.

Figure 4.7 shows the removal rate of Cu and TaN films versus the PAM concentration. The removal rate of Cu film slightly decreased with PAM concentrations. Here, the removal rate of TaN film was strongly suppressed and saturated with increasing PAM concentrations in aqueous suspension as shown in Figure 4.7.

To enhance the removal selectivity of Cu-to-TaN films with suppressing the removal rate of TaN film by selective adsorption, we also optionally added organic polymer (PAM) with the concentration of up to 0.7 wt%. The adsorption of PAM-added slurry on the abrasive particles and the film surfaces corresponds to the differing zeta potential charge. By this zeta potential difference, the removal rate of the Cu and TaN films was more suppressed, and the oxide-to-nitride removal selectivity increased with addition of PAM.

Figure 4.8 shows the electrokinetic behaviors of Cu film, TaN film, and colloidal silica slurries with PAM addition as a function of pH.

Adsorption of PAM on Cu and TaN film surfaces increases and reaches a strong suppressed point of approximately 0.3 wt%. In other words, PAM is more adsorbed on the Cu and TaN film surfaces. This is driven by the difference in zeta potential, which affects the interaction between PAM and each surface.

In addition, above the isotropic point, the slightly negative-charged Cu oxide and TaN films surface can interact with the deprotonated between carboxyl groups of alanine, neutral $-NH_2$ groups, and NH+ functional groups of PAM, which results in the formation of strong complexes with Cu and TaN films. However, with addition of PAM, the removal rate of

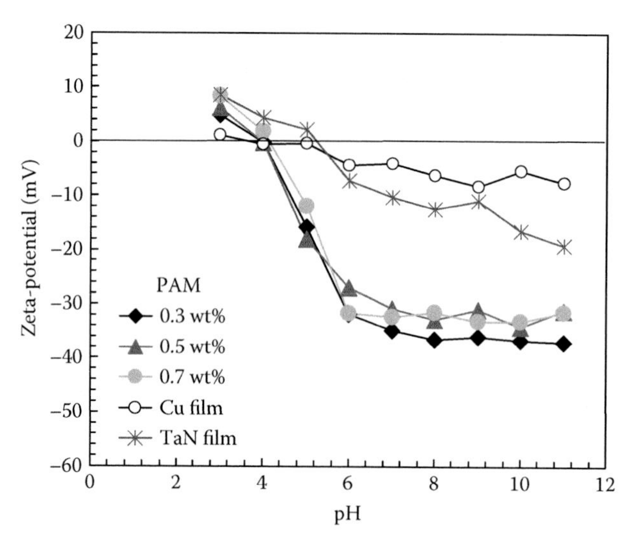

FIGURE 4.8 Zeta potential of Cu, TaN films, and colloidal silica slurry with PAM as a function of pH.

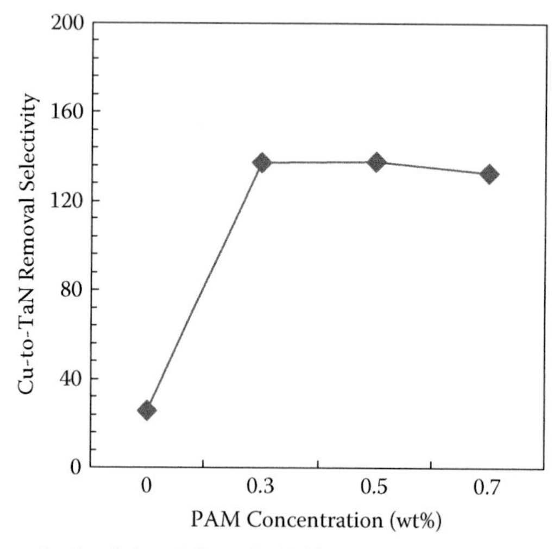

FIGURE 4.9 Removal selectivity of Cu-to-TaN films versus PAM concentration in slurry.

Cu and TaN film decreased. By increasing the PAM concentration, the removal selectivity drastically increased and essentially saturated from 30:1 to 130:1 (Figure 4.9).

Potentiodynamic polarization studies were carried out to measure the corrosion current density and potential at various alanine and polyacrylamide concentrations with H_2O_2. Polarization plots for Cu film as a function of alanine concentration with H_2O_2 at pH 10 are presented in Figure 4.10a and b, respectively. The value of corrosion potential reduced gradually

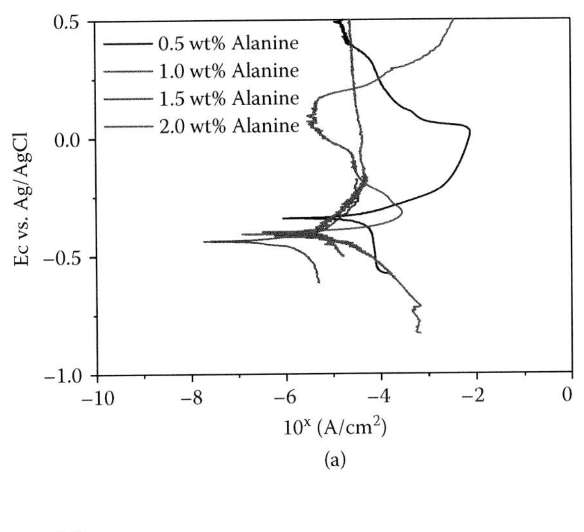

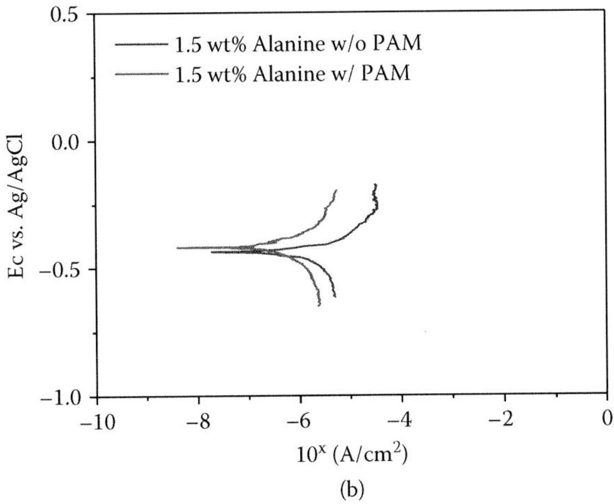

FIGURE 4.10 (See color insert) Potentiodynamic polarization: (a) various alanine concentration, (b) 1.5 wt% alanine with and without PAM.

with an increasing concentration of alanine (Figure 4.10a). As the alanine solution increases from 0.5 to 2.0, the fraction of carboxyl and amino functional group is more pronounced with increased anion fraction at alkaline pH. Since the bidentate (L–) is more reactive than the monodentate (HL), the increased dissolution rate of Cu^{+} or Cu^{2+} ions. The removal rate of Cu film increases with an increasing alanine concentration by low corrosion potential. On the other hand, corrosion potential showed no difference with addition of PAM solution (Figure 4.10b).

4.3 Copper CMP Pattern Dependence

The use of copper as an interconnect material in multilevel scheme DRAM and NAND flash memory in design rules beyond 45 nm is being increasingly considered mainly due to copper's low resistivity and high resistance to electromigration compared to the widely used aluminum (Al) alloys. As the device structure becomes more complicated and its dimensions shrink, the conductors on the chip must be thin enough to occupy less space. Such miniaturization of the conductor causes an increase in the RC delay, which is the product of the metal resistance (R) and the capacitance (C) of the interlevel dielectric. While the low resistivity of copper is expected to exhibit lower interconnect delay, high resistance to electromigration enhances the device reliability by increasing the mean time to device failure. In addition, in the manufacture of memory circuit devices with copper metallization, multilevel interconnects are formed using the damascene method, whereby copper is deposited by chemical vapor deposition (CVD) or electroplating into vias and trenches etched in the interlayer dielectric (ILD) over a diffusion barrier usually made from titanium (Ti), tantalum (Ta), or their nitrides. For the application of copper as interconnect material, the film surface within wafer must be made planar on a global scale. Inlaid metal patterns in multilevel chips could be obtained by CMP. The CMP is used to planarize the barrier metal, low-k, and copper layer following their deposition process.

4.3.1 Dishing Dependency on Feature Size and Pattern Density

Dishing of copper lines is among the most important issues of copper CMP. Dishing reduces the final thickness of copper lines and degrades the planarity of the wafer's surface, resulting in complications when adding multiple levels of metal. Understanding of dishing and its nature is helpful in process optimization and in understanding the process mechanism. Here, we present a thorough investigation of dishing in copper CMP. Along with studying the dependency of dishing on linewidth and pattern density, our investigation is focused on the effect of (over)polish time, oxidizer concentration in the slurry, and thickness of the as-deposited copper layer. As a result, a hypothesis of material removal mechanism for our type of slurry is presented.

The test structures were fabricated as following. First, the interconnection grooves were etched in thermally grown silicon dioxide by RIE with the depth of 600 nm. The width of the trenches varies from 2 to 100 mm. Second, after depositing a 50-nm Ti layer by sputtering an adhesion promoter layer, an 800-nm Cu film was deposited also by sputtering without breaking the vacuum. An IC1000/Suba IV stacked-perforated pad from Rodel was applied. Prior to every run, pad conditioning was done using diamond tool. An alumina-based slurry (Al_2O_3 particles, median diameter 200 nm), which contains a complexing (carboxylic acid) agent and a stabilizer, was used throughout the investigation. Hydroperoxide (functions as oxidizer) was added to the slurry at three volume percentages: 7.5%, 10%, and 15%. After mixing, the pH of the slurry was measured to be around 4.0. Polishing runs were carried out using a Presi Mecapol E460 polishing tool. Optimum settings for uniformity were applied, which were found for our test pattern. The polishing pressure was 250 g/cm^2. The rotation speed of platen/pad and that of the wafer holder were set at 50 rpm. The supply speed of the slurry was 125 ml/min. Due to practical limitations, we used a timed polish process. The nominal polish time was determined by the moment when the entire wafer surface is clear from excess metal. From then on, if the wafer polishing is continued, we consider it as overpolishing and the time is called overpolish time. In our study, four polish times were used, which are nominal, 5% (of nominal time) overpolish, and 10% and 20% overpolish.

Figure 4.11 shows the dishing amount of copper lines at different linewidth and pattern density. The wafer was polished using slurry with 15% hydroperoxide and the polish time was nominal. As expected, the dishing amount strongly increases with the increment of linewidth (see Figure 4.11). The dependency is not a linear function of linewidth. At a linewidth above 50 mm, the dishing levels off. The pattern density only shows a minor effect on copper dishing. Unlike dishing data published for other material, such as tungsten, dishing of copper lines appears to be relatively large even at nominal polishing. As shown in Figure 4.11, the dishing amount of 100-mm wide copper lines is more than 100 nm. However, it is well known that the removal rate is higher at dense areas, thus when the entire wafer surface is clear, which we defined as nominal polishing, dense areas must have been overpolished. In addition, there is always a certain non-uniformity of removal rate over the wafer. Therefore, when the entire wafer surface is cleared there definitely are areas that have been overpolished. We assume that overpolishing with very high removal rate of copper (typical 600 nm/min), high selectivity between copper and ILD (typical larger than 90), and a too thin as-deposited copper layer are the reasons of the large amount of dishing.

Overpolishing is needed to ensure good electrical properties of interconnection (no shorts between separate interconnect). However, overpolishing always results in an increasing amount of dishing and worsens the planarity of the wafer surface. Figure 4.12 shows the profiles of a test structure of 20-mm wide copper lines with a pattern density of 50% at nominal polish time and three overpolish times. The dishing increases dramatically with increasing overpolish time. As many authors have described, dishing as a result of the pad reaching into recess areas and removal of copper in the recess, there is a question raised if the dishing rate is the same at different linewidths. Therefore, we plotted the dishing

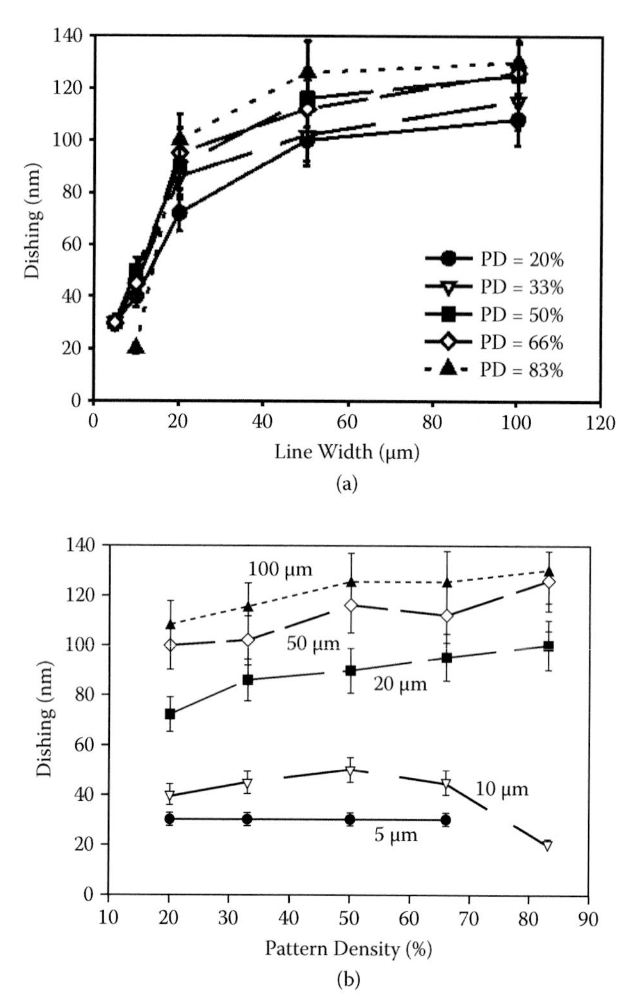

FIGURE 4.11 Dishing versus (a) linewidth and (b) pattern density.

rates of different linewidths versus overpolishing time in Figure 4.13. The dishing rate of wide lines is higher than that of narrow ones. To explain this, we use the model proposed by Warnock saying that the pad reaches into recessed areas by bending and its roughness. The amount that the pad can reach into the recessed areas depends on the pad's properties (e.g., hardness, surface roughness), linewidth, and applied pressure. Since all the other conditions remain the same in our case, linewidth is the only factor that can affect the amount of pad reaching. Therefore, it directly relates to the amount of dishing as well as to the dishing rate. The model is thus in accordance with the obtained dishing rate behavior seen in Figure 4.13, that is, the different slopes are explained.

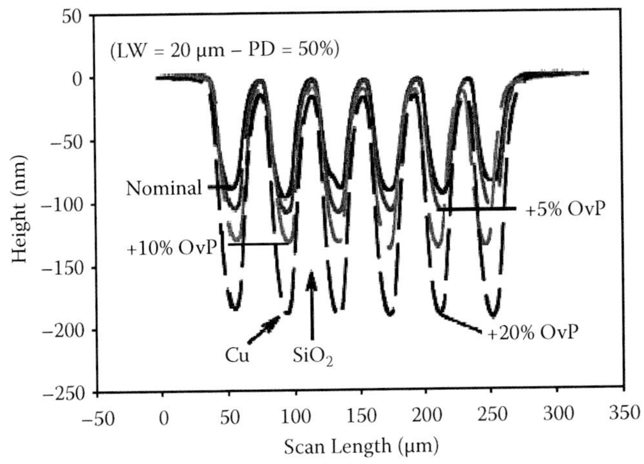

FIGURE 4.12 Surface profiles of a polished structure (20 μm linewidth, PD = 50%) at different polish times.

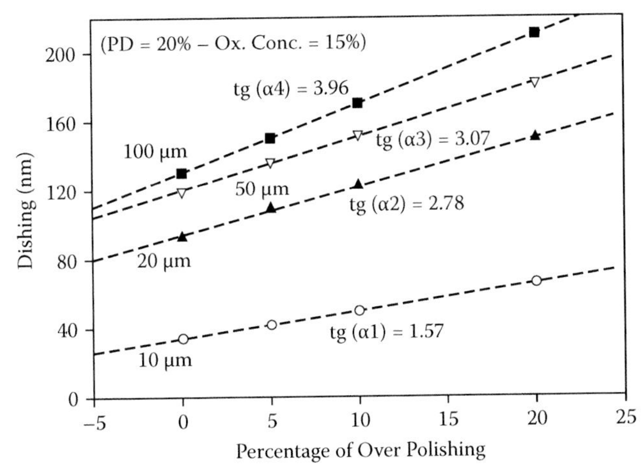

FIGURE 4.13 Dishing of different linewidths at different polish times.

Slurry chemistry has been reported to have a strong effect on polishing results. We also found a strong dependency of dishing on slurry chemistry. Figure 4.14 illustrates the dishing of 100-, 50-, and 20-mm wide lines (pattern density 20%) polished by slurries with different concentrations of hydroperoxide at nominal polish time. It can be seen that the dishing is reversely proportional to the concentration of oxidizer in the slurry. The explanation we propose for this phenomenon is that with higher oxidizer concentration in the slurry, a more effective passivation layer is formed on the copper surface (it will grow faster). This passivation layer slows the removal rate of copper in the recess areas and better protects the copper lines from dishing during the overpolish step. We have found that the

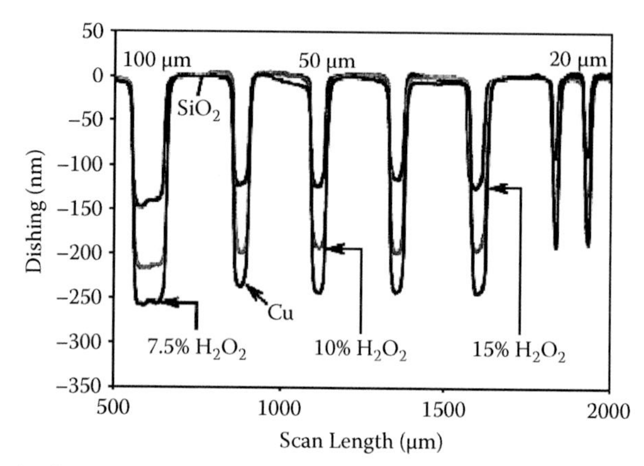

FIGURE 4.14 Surface profiles of a structure (PD = 20%) polished by slurry with different oxidizer concentrations.

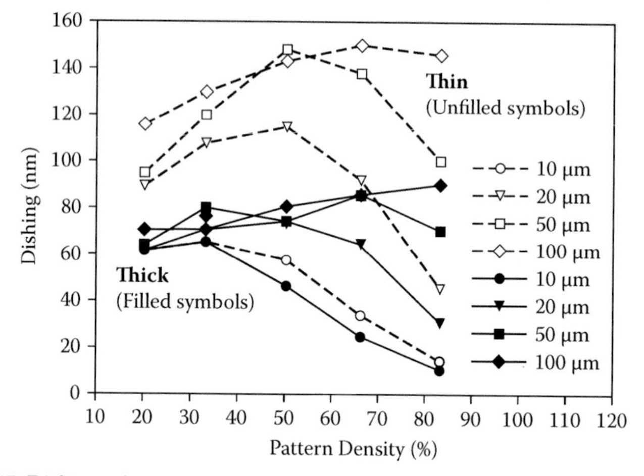

FIGURE 4.15 Dishing of as-deposited thin and thick copper layers at different linewidths and pattern densities.

thickness of the as-deposited copper layer also affects the amount of dishing of copper lines. Two copper thicknesses are used to study this dependency. The thin copper layers with a thickness of 800 nm was described earlier. The thick copper layer is 1.5 mm thick. Dishing data for both thicknesses at nominal polish times are shown in Figure 4.15. It is clear that the amount of dishing is smaller in all cases for the as-deposited thicker copper layers. Furthermore, the effect of thin versus thick copper layer on dishing appears to be even clearer at the large features, while only minor effects are observed for the small features. It is known that the removal rate of copper at dense areas is higher than that of field areas due to higher local pressure exerted on the features. Therefore, if the copper layer is too thin, the dense areas will be cleared first and experiences overpolishing before the field areas are cleared and global planarity is reached. This leads to severe dishing at the dense areas. The copper layer should be sufficiently thick to reach a globally planar surface with still excess of copper on all features (wide, small, densely and widely packed). This will minimize the dishing. In this way, we managed to reduce the maximum dishing for all investigated feature sizes to less than 100 nm for nominal polishing conditions.

Figure 4.16 shows etch rates of copper in the slurry without abrasive particles at three investigated concentrations of oxidizer. We notice that the static etch rate of copper is very low (10 nm/min). On the other hand, the removal rate of copper during polishing is very high (more than 600 nm/min is achieved). This is strong evidence of the forming of a passivation layer on the copper surface in the slurry. Further investigation of the chemistry of the slurry gave us the following hypothesis about the forming of the passivation layer on the copper surface and thereafter a proposal for the copper removal mechanism in copper CMP with our slurry. The oxidizer (H_2O_2) reacts with Cu in acidic slurry (pH 4) and Cu^{2+} ions are formed. The anions of the carboxylic acid react with Cu^{2+} ions ($R(COO)_2Cu$). Carboxylates of metals other than the alkali metals generally are insoluble. Therefore, we suppose that $R(COO)_2Cu$ protects the copper underneath from etching. According to the Pourbaix diagram for the copper–H_2O system, no copper oxide can be formed in our slurry (the pH of our slurry is about 4.0; at this pH, only two forms of Cu^{2+} or Cu are possible; see Figure 4.17). The concentration of H_2O_2 strongly influences the amount of Cu^{2+} ions and, therefore, the amount of $R(COO)_2Cu$ product. In other words, the effectiveness of the passivation layer is directly proportional to the H_2O_2 concentration in the slurry. This is consistent with our experimental results.

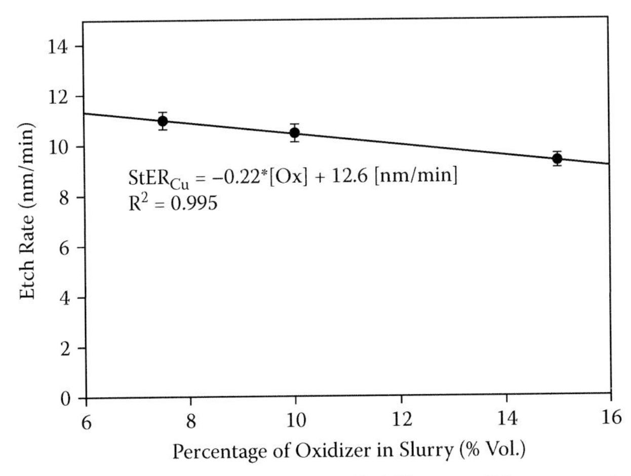

FIGURE 4.16 Etch rate of copper in the slurry with different oxidizer concentrations.

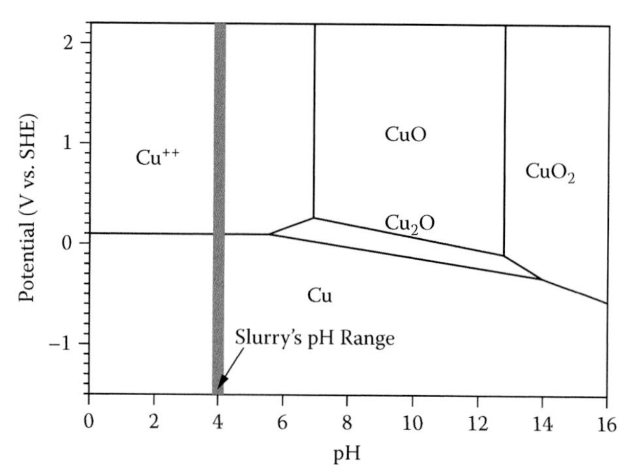

FIGURE 4.17 Pourbaix diagram of the copper–H_2O system.

The hypothesis for the copper removal mechanism using this slurry is proposed to be as follows. Copper on the surface is oxidized by H_2O_2 (in acidic environment) into copper cations. These cations then react with carboxylate anions to form the passivation layer that protects copper from etching. This layer is then removed at protruding levels by mechanical abrasion. Once removed from the surface, the "metallic soap" particles are swept away by the turbulent motion of the slurry. Further investigations of the passivation mechanism by varying the slurry chemistry are being conducted.

The dishing behavior of copper lines has been extensively studied. Relationships between dishing and feature size, pattern density, overpolishing time, thickness of as-deposited copper layer, and slurry chemistry have been elucidated. While dishing strongly depends on linewidth, only a small effect of pattern density has been observed. As expected, dishing dramatically increases with the increment of overpolishing time. Furthermore, the dishing rate dependency of overpolish time is not the same for all feature sizes. The larger the linewidth, the higher the dishing rate is. Thick as-deposited copper layers yield less dishing than thinner ones. The oxidizer concentration in the slurry also has a strong impact on the amount of dishing. It was found that, within the investigation window, the dishing is inversely proportional to the oxidizer concentration. From the obtained results, a hypothesis for the passivation layer formation has been proposed and the mechanism of copper removal has been presented.

4.3.2 Pattern Effects on Planarization Efficiency of Cu Electropolishing

Cu electropolishing technology has been explored as a replacement of the Cu CMP planarization process. Contolini et al. (1994) integrated Cu electropolishing with a wet etching technology in a novel electrochemical planarization (ECP) method for Cu multilevel interconnects, and Wang (2000) has designed a commercial tool for the electropolishing process. In addition, Cheung (1997) proposed a process of Cu electropolishing to enhance CMP throughput. Recently, groups at TSMC and SONY companies also applied Cu electropolishing in global planarization technologies.

Traditional electropolishing is an important surface treatment technology, and can dissolve a metallic film uniformly and produce a smooth and bright surface. Furthermore, electropolishing has potential advantages in that it renders a reduced waste stream, is less consumable, and there is no applied pressure to the substrate, which is beneficial for future low-dielectric-constant-material integrated processes. In previous studies of Cu CMP, Steigerwald et al. (1997) found that Cu dishing is a strong function of linewidth but is only weakly dependent upon pattern density. At the same pattern density, the amount of dishing increased as the linewidth increased. In this work, pattern effects of Cu electropolishing were discussed. Anodic potentiodynamic polarization measurement was also employed to clarify the dissolution mechanism of Cu electropolishing.

The patterned wafer used was composed of a 30-nm-thick ionized metal plasma (IMP)-TaN layer as the diffusion barrier, and a 200-nm-thick IMP-Cu film as the seed layer. The experiments on Cu electroplating and electropolishing were carried out in a tank of nonconducting material at room temperature. The counterelectrode was a platinum plate and the working electrode was a sliced wafer with a size of $2 \times 3\ cm^2$. In Cu electroplating, the electrolytes included $CuSO_4 \cdot 5H_2O$ (30 g/L), H_2SO_4 (275 g/ L), chloride ions (50–100 ppm), polyethylene glycol (40–2000 ppm), and 2-aminobenzothiazole (10–100 ppm). The films were deposited under galvanostatic control. In Cu electropolishing, the electrolyte was phosphoric acid (H_3PO_4) and the films were polished under potentiostatic control. Potentiodynamic (PD) polarization measurement was performed on an EG&G potentiostat/galvanostat (model 273A) with a Pentium PC. In these analytical experiments, the counterelectrode was platinum and the working electrode was Cu with a constant surface area of $0.5\ cm^2$. All potentials are reported relative to the Ag/AgCl electrode, which was used as the reference electrode. Cross-sectional profiles of Cu films were examined using a field emission scanning electron microscope (FESEM). Surface roughness was measured using an atomic force microscope (AFM). The sheet resistance of Cu deposits was measured by the four-point probe technique, and the resistivity measurements were carried out immediately after deposition.

Cu planarization process using ECP of Cu by electropolishing followed by CMP is depicted in Figure 4.18. After Cu electroplating completely fills the trenches and vias, electropolishing planarizes the surface down to the barrier layer, and the remaining Cu and the barrier metal are removed

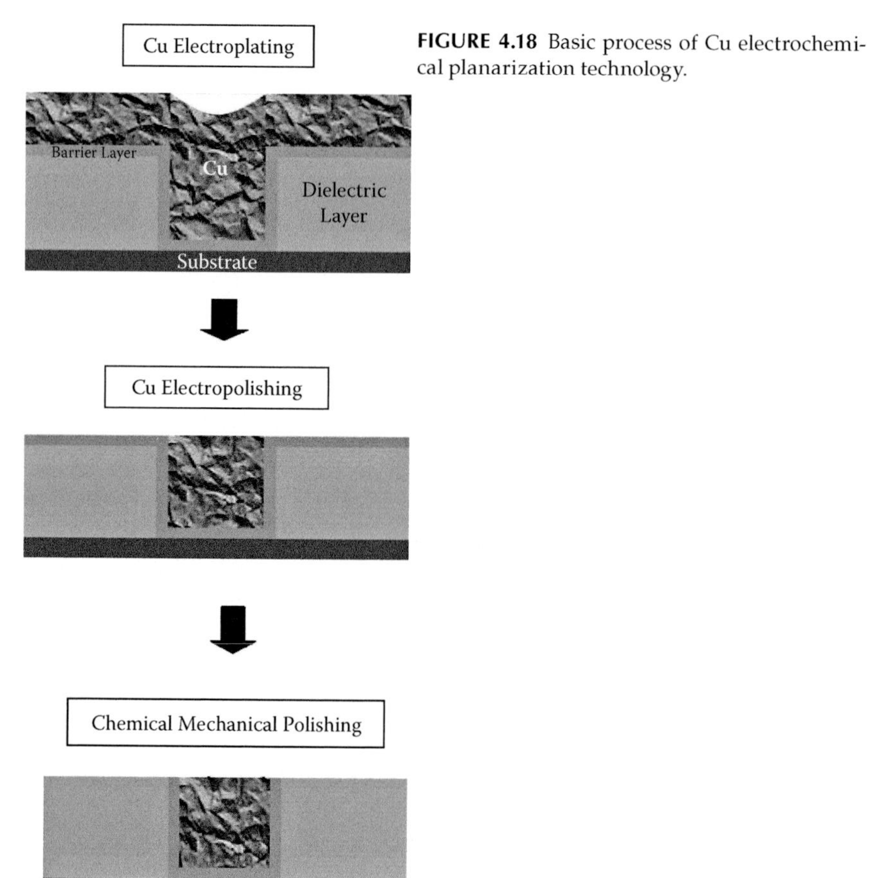

FIGURE 4.18 Basic process of Cu electrochemical planarization technology.

by a single-step CMP process. Figure 4.19 shows the scanning electron microscope (SEM) cross-sectional profile of a 10-μm Cu line planarized by the ECP process. In this case, the defect-free filling was obtained by an acid cupric sulfate electrolyte containing chloride (Cl), polyethylene glycol (PEG), and 2-aminobenzothiazole (2ABT). Subsequently, a clean and scratch-free surface was produced by electropolishing using H_3PO_4 as the electrolyte. Finally, CMP removed the remaining Cu and the barrier metal. The final-step CMP process used the H_2O_2-based slurry, which has a higher removal rate of TaN than that of Cu. For Cu electroplating, the combined action of Cl–PEG–2ABT provided an inhibition gradient between the opening and the bottom of a feature to obtain an obviously selective deposition and to result in bottom-up filling. Tafel plots in Figure 4.20 reveal that the added 2ABT could enhance the charge transfer resistance to inhibit Cu deposition. The shifted overpotential, caused by the added PEG, was 61.7 mV relative to that of standard solution and the corresponding value for the combined action of PEG–2ABT was 77.2 mV.

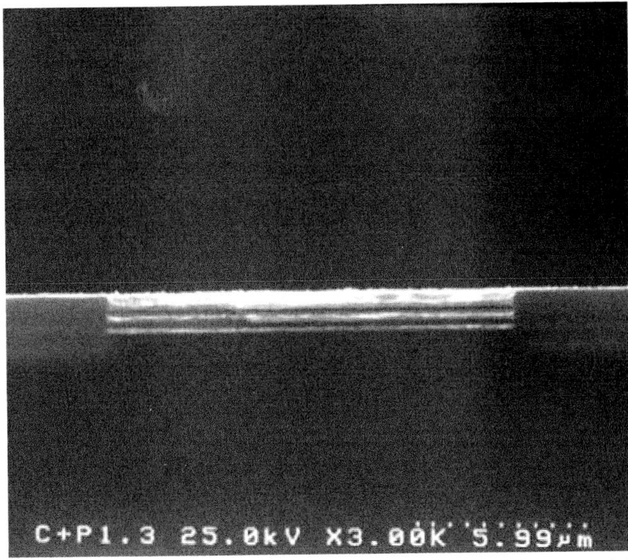

FIGURE 4.19 SEM cross-sectional profile of a 10-µm Cu line planarized by the ECP process.

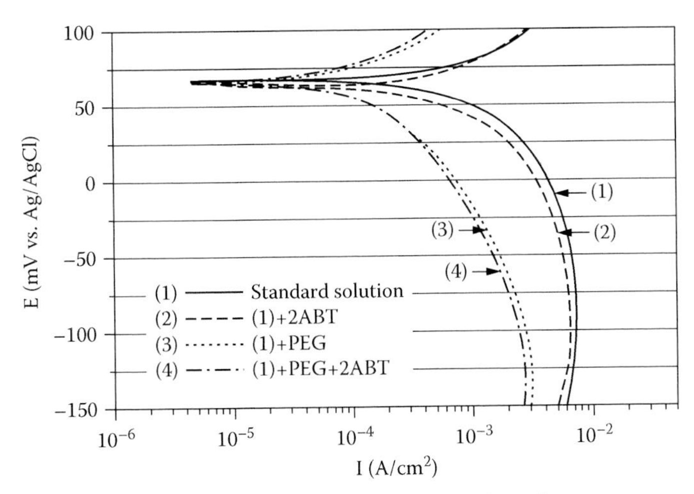

FIGURE 4.20 Tafel plots of Cu electroplating with various electrolytes.

The potentiodynamic curve of Cu electropolishing in the H_3PO_4 (85%) solution is shown in Figure 4.21. For anodic potentials in the AB range, the metal surface became active. When the anodic potential is higher than the B point to the BC range, a viscous sublayer may start to form on the anodic surface. In the CD range, called the plateau region, a wide passivation range existed; electropolishing occurred with negligible change in current density as the applied voltage increased. In this plateau operation

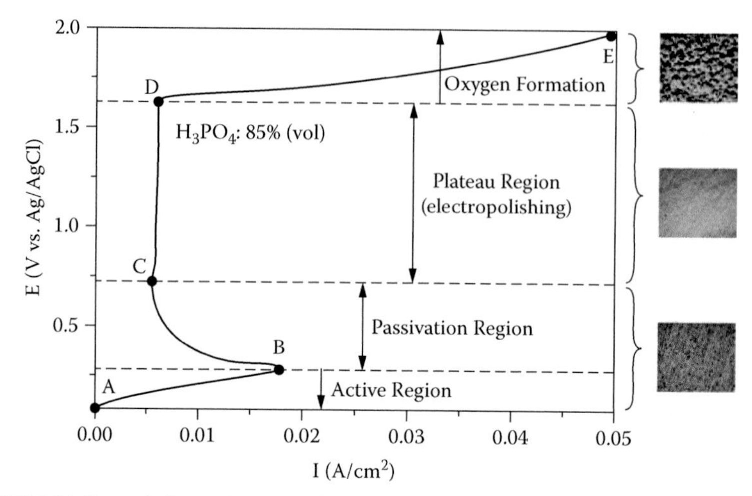

FIGURE 4.21 Potentiodynamic curve of Cu electropolishing. In addition, the optical microscope images (200×) show surface morphologies of Cu electropolished at different applied voltages.

region, a nearly constant current maintained a stable electropolishing process. Unlike the cathodic-limiting-current plateau in electrodeposition, which results from the depletion of metal ions in the diffusion layer near the electrode surface, the formation of anodic-limiting-current plateaus in electropolishing may be due to the presence of a viscous film on the anodic surface or the concentration barrier formed by accumulated dissolving metal ions.

In the DE range of Figure 4.21, the electropolishing process occurred quickly because of the high applied voltage. An increase in current in this stage increased the rate of oxygen formation from the breakdown of water in the electrolyte with increasing potential. This oxygen formation reaction caused severe etch pits to be formed on the Cu surface. Optical microscope images in Figure 4.21 show that at applied voltages lower than point C, the polished surface was slightly etched; when the potential was in the plateau region (in the CD region), a smoother and scratch-free surface was obtained. Furthermore, when a high potential was applied (in the DE region), a greater amount of oxygen bubbles was generated and the working electrode surface was pitted.

In the process of Cu electropolishing, the polishing rate was constant and determined by measuring the remaining Cu thickness of the blanket wafer with a 1-μm-thick Cu film. The polishing rates of electroplated Cu films were about 500 nm/min, 1000 nm/min, and 1500 nm/min for 85% (vol.), 70% (vol.), and 50% (vol.) of H_3PO_4 electrolytes, respectively, as calculated from the data in Figure 4.22. The fluctuation of polishing rates away from linear fitting may be due to non-uniform current distribution on the residual Cu film. After electropolishing, the average roughness (Ra) of the

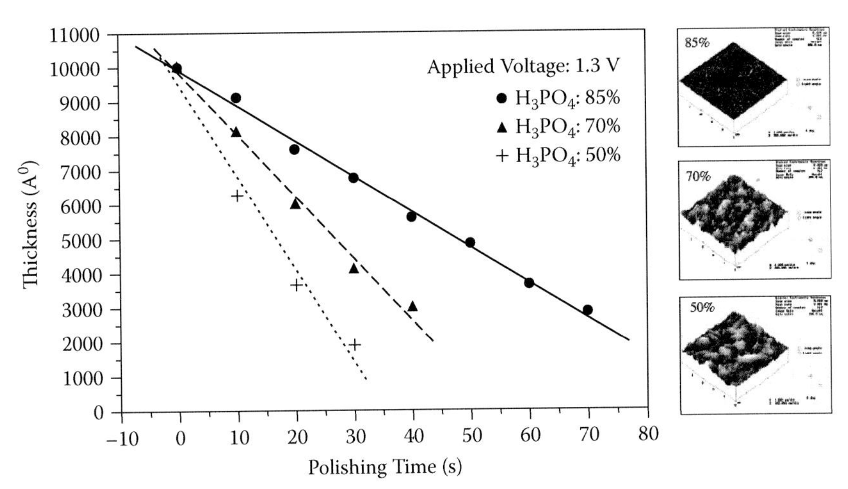

FIGURE 4.22 Effects of acid concentration on polishing rates and surface morphologies (AFM images) for Cu electropolishing at applied voltage of 1.3 V (with respect to the Ag/AgCl reference electrode).

films decreased with increasing acid concentration, as seen in Figure 4.22. Therefore, electropolishing is normally carried out in a limiting-current region and in a concentrated acid electrolyte; a higher acid concentration showed a higher leveling effect.

Generally, conventional Cu electroplating will produce step on/between features. The traditional Cu CMP process includes step-height reduction of wiring metal and removal of overburden metal outside the features. Steigerwald et al. found that Cu dishing is a strong function of linewidth, but is only weakly dependent upon pattern density. At the same pattern density, the amount of dishing increased as the linewidth increased. In this study, we also encountered the same issue for Cu electropolishing. To measure the planarization efficiency (PE) of the CMP process, Steigerwald et al. defined the following equation:

$$PE = [1 - (\Delta down/\Delta up)] * 100\% \tag{4.3}$$

where down and up are the thickness differences of the inside and outside of the feature respectively, as shown in Figure 4.23. In this article, we also applied Equation 4.3 is also applied to monitor PE of Cu electropolishing. A better planarization ability is noted when the PE value is higher. Ideally, PE is equal to 1. The following mechanism of electropolishing is suggested: the microleveling effect occurs because of selective dissolution. When current is applied, a passivation film covering the crevices of the surface—which has a high specific gravity, viscosity, and insulation—prevents dissolution; whereas the surface protuberances not covered by the passivation film—which receive greater current from the

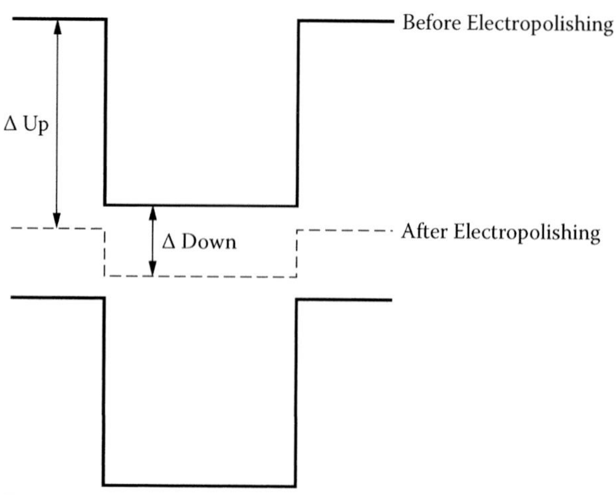

FIGURE 4.23 Diagram of PE measurements, where Δdown and Δup are the thickness differences inside and outside of the feature, respectively.

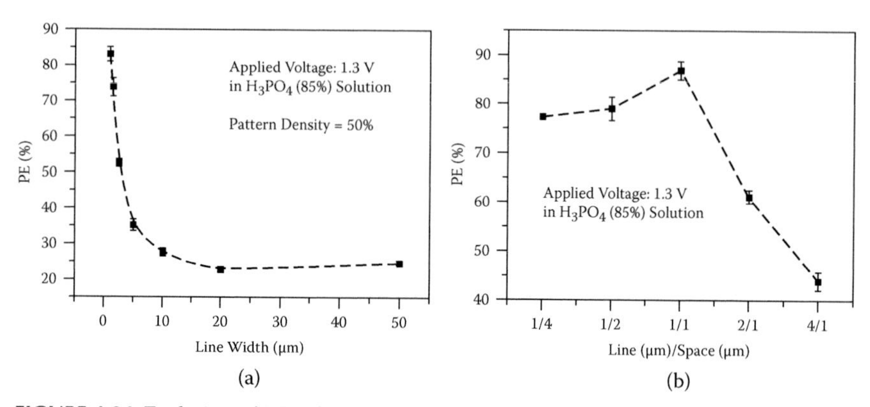

FIGURE 4.24 Evolution of PE values with (a) different linewidth and (b) different ratio of linewidth to space width.

cathode—dissolve more quickly. This phenomenon leads to a lower dissolution gradient in larger features such as in a blanket surface. Therefore, Figure 4.24a shows that PE decreased with increasing linewidths when pattern density [line/(line + space)] remained at 50%. By sputtering Cu into the filled features, starting profiles of filled features with greater step height manifest the pattern effect on planarization efficiency.

The SEM images in Figure 4.25 indicate that the capability of step-height reduction in small patterns was higher than that in larger patterns. In these cases, the polishing time was 150 s. Table 4.1 summarizes the starting and final (after electropolishing) step heights of different features. Moreover, PE decreased more quickly for narrower lines than for wider lines. In addition to the linewidth dependence of PE, Figure 4.24b also

TABLE 4.1
Starting and Final (after Electropolishing) Step Heights of Different Features

Line/Space (μm)	Starting Step Height (nm)	Final Step Height (nm)
1/1	~850	~480
2/2	~730	~420
5/5	~720	~530
10/10	~720	~550
20/20	~720	~550
1/4	~1300	~700
1/2	~1300	~640
2/1	~850	~360
4/1	~760	~260

shows that PE decreased with an increase in space width when linewidth remained unchanged. We suggest that dissolution current density in a smaller space was higher than that in a larger space, thereby resulting in a higher polishing rate around the outside of the feature or leading to an enhancement of the dissolution gradient between the gap and spacing. However, the influence of space width on PE was lower than that of linewidth, as shown in Figure 4.24b.

In these Cu electroplating experiments, only PEG and inhibitor were used, so there was no overplating but dishing occurred with about a 100 nm step height in a 10-μm Cu line. Cu electropolishing was capable of eliminating the step height of such an electroplated Cu line, as shown in Figure 4.19. However, overplating has recently been observed for a conventional bath with a brightener. Nevertheless, Cu electropolishing is still able to yield a planar surface due to a higher polishing rate for overplated protrusions with higher current density than that for blanket regions

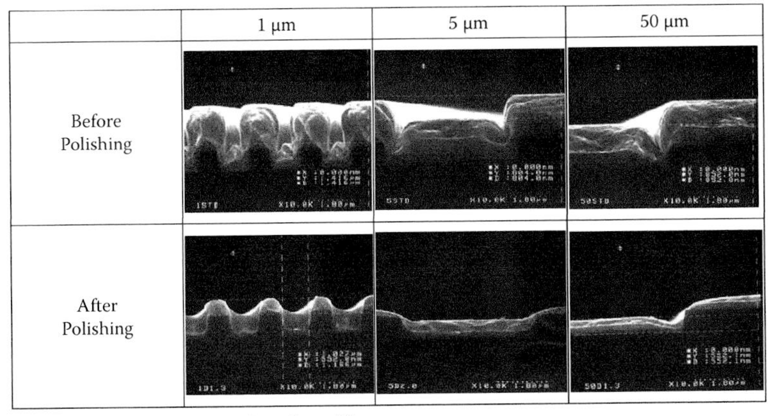

FIGURE 4.25 SEM cross-sectional profiles.

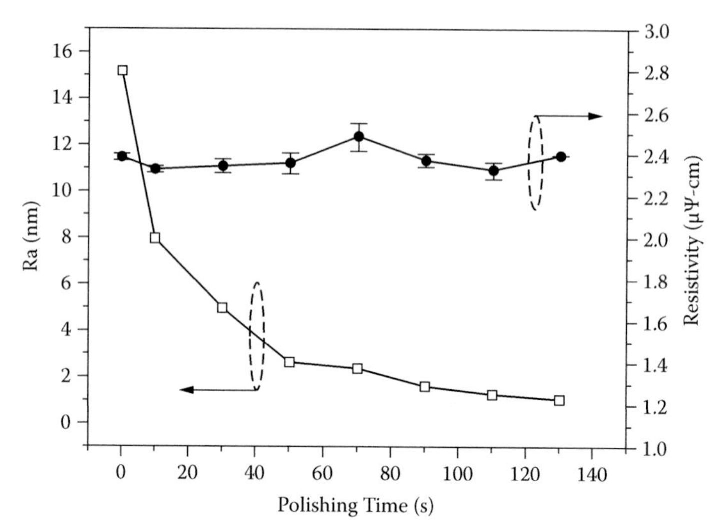

FIGURE 4.26 Average roughness (Ra) and resistivity of blanket Cu films with various polishing times.

with lower current density. On the other hand, non-uniform current distribution is a challenge in the case of Cu electropolishing for global planarization of an entire wafer. Adding additives into polishing baths or introducing a multistep pulse current could improve the global uniformity within a wafer.

Figure 4.26 reveals that the average roughness (Ra) of blanket Cu films decreased as polishing time increased. At the beginning of electropolishing, the microleveling effect was more obvious than at later times. The trend in this figure mainly followed from the fact that a point-discharge effect is more efficient for a rough surface than for a polished surface. After electropolishing for 130 s, Ra was approximately 1.1 nm, as compared to 13.2 nm before electropolishing. The resistivity of polished films was not obviously changed, as also shown in Figure 4.26. This result suggests that no H_3PO_4 electrolyte diffused into grain boundaries of Cu films, whereas some oxidants often cause such damage after CMP processes, thus degrading the electrical characteristics of polished Cu metals.

4.3.3 Cu Pad Size and Linewidth Affect Dishing

Although the exposed Cu can immediately react with oxygen to form an oxide film, the film is porous and not of a self-protective nature. Therefore, a capping material, such as SiN, is necessary to prevent the corrosion of Cu. Unfortunately, the Cu corrosion depends on the delay time from the CMP polish of Cu to deposition of the protective layer. On the production line, the manufacture available time and efficiency are very hard to reach these stiff conditions. Hence, the prevention of Cu corrosion becomes the major challenge in production to improve the yield. On the other hand, the grains of Cu will enlarge during subsequent thermal processing. For instance, the heating during dielectric material deposition would induce the Cu grain growth and generate the voids in Cu due to the surface area diminution of grain boundary. This will result in poor thermal stability of Cu film. In the Cu CMP process, dishing and erosion are the other problems for the Cu line thickness control. The occurrence of dishing and erosion depends on the line width and density. Besides, the CMP parameters such as polishing down force, polish head rotation speed, polish pad elastic properties, slurry flow, polish time, and so forth affect dishing and erosion. Capping with a protective layer is the plausible way to avoid the dishing and erosion phenomena so as to improve the Cu thermal stability. This section investigated the dishing and erosion phenomena of Cu in the CMP process. The tantalum nitride (TaN) capping on the top of Cu surface is proposed to protect the Cu from corrosion and oxidation. The thermal property of Cu is also examined using stress migration to evaluate its stability.

A three-metal-level Cu interconnect was performed. In metal 1, a single damascene structure was applied. Dual damascene structure of via 1, 2 and metal 2, 3 was applied. IMP sputtering of TaN was utilized for Cu barrier deposition and self-ionized plasma sputtering of Cu was applied for the seed layer deposition of electroplating Cu. Overburden Cu was polished in a linear system. A two-step polish (copper and barrier metal were separated polish) system was carried out for the Cu-CMP process. An aluminum oxide abrasive system was selected for both of the polishing steps. A TaN capping process was carried out after CMP polishing. The TaN of 30 nm in thickness was deposited and repolished away at the second step of Cu CMP (Figure 4.27). Because of the selectivity effect, there remained a very thin TaN layer on the Cu surface. The Tencor HRP-20 microprofiler measured dishing and erosion. The thermal stability of Cu was evaluated in a furnace at 180°C for 170 h. The resistance of metal, *Rs*, was measured based on a serpentine test structure and the resistance

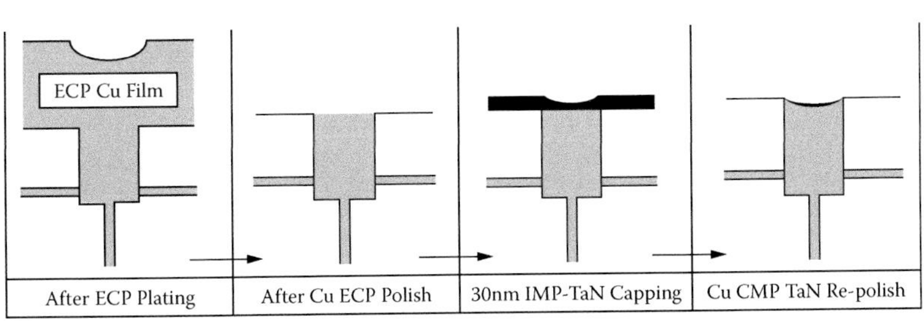

FIGURE 4.27 TaN capping process procedure. 30 nm TaN was capped after Cu CMP, and after TaN repolishing there remains a very thin TaN layer on the Cu surface.

of via, *Rc*, was measured based on via chain (totally 1798 via embedded with 2 mm in length and 1.5 mm in width) structure. A KLA-Tencor AIT-II was used to evaluate the defects induced by Cu corrosion.

4.3.3.1 Pattern Dependence of Dishing and Erosion Phenomena

The influence of pattern density on dishing and erosion was characterized and the results are shown in Figure 4.28 and Figure 4.29, respectively. Figure 4.28 illustrates that the Cu pad size and linewidth affect the dishing. As the Cu pad size increases four times, the dishing increases

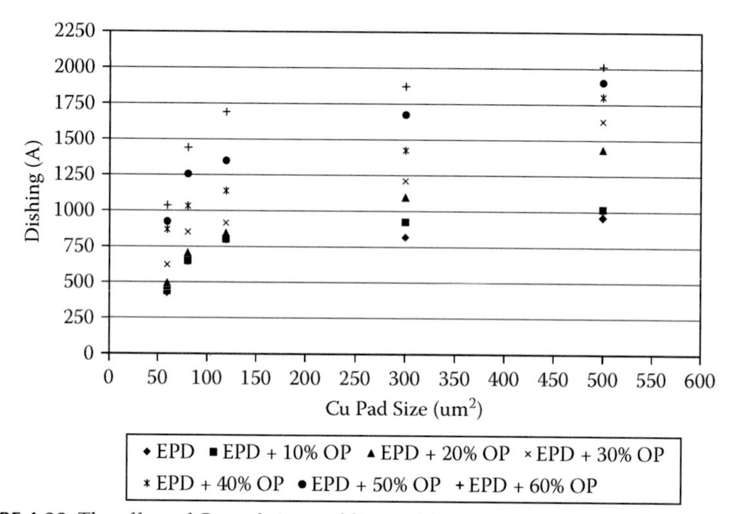

FIGURE 4.28 The effect of Cu pad size and linewidth on dishing.

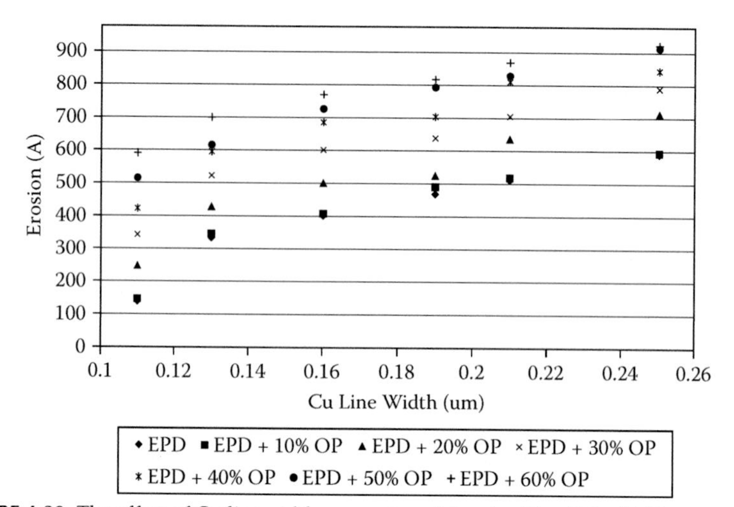

FIGURE 4.29 The effect of Cu linewidth on erosion at fixed oxide pitch of 120 µm.

TABLE 4.2

Selectivity of Different Cu Polish Modules

CMP Polish Module	Cu	TaN	FSG	Thermal Ox
1st module for Cu polish	115	4	0.7	1
2nd module for TaN polish	1	5.8	3	2.3

to approximately two times. The erosion exhibits the same trend; it increases with the linewidth at a fixed oxide spacing of 120 mm, as shown in Figure 4.29.

In Figure 4.28, the dishing increases in a very fast rate at the small sizes of Cu pad area ranging from 60 to 150 mm^2. In small Cu pads, the oxide plays a very important role in antidishing. The higher the ratio of oxide area, the lesser the dishing observed. This results from the high selectivity of oxide in the Cu polishing module using the aluminum slurry system. The selectivity of different Cu CMP polishing modules is shown in Table 4.2.

On the other hand, the percentage of Cu overpolish will worsen the result. As the percentages of overpolish increase, the total polish time increases. This, in turn, immerses the wafer in the slurry for longer times. The additional immersion time caused Cu corrosion by slurry chemical reaction. This explains why more overpolish causes a higher degree of dishing. For the Cu pads larger than 150 mm^2, the dishing increased at a nearly constant rate. In these cases, the CMP polish pad deformation dominated because the CMP polish pad is made of polyurethane that would be deformed during polishing and provide the CMP planarization. The limitation of polish pads deformation causes the constant increasing of dishing. As shown in Figure 4.29, the erosion also increases at a constant rate. Fixed densities of Cu lines and oxide pitch (120 mm) are the major cause of this phenomenon. Because the selectivity of Cu to oxide is approximately 200:1 (Table 4.2), the larger oxide area will reduce the erosion amount in small line width. The increase of erosion only depends on the Cu linewidth at fixed oxide pitch. Besides, the higher overpolish exhibited the same behavior, as observed in dishing experiments; the slurry chemistry effect is again dominant.

4.3.3.2 TaN Cap Process for Cu Corrosion Prevention and Thermal Stability Improvement

The dishing and erosion generated by the Cu CMP process could be controlled by an appropriate Cu-to-oxide-area ratio. After Cu polishing, an IMP–TaN layer was sputtered onto the wafer surface. The overburden TaN

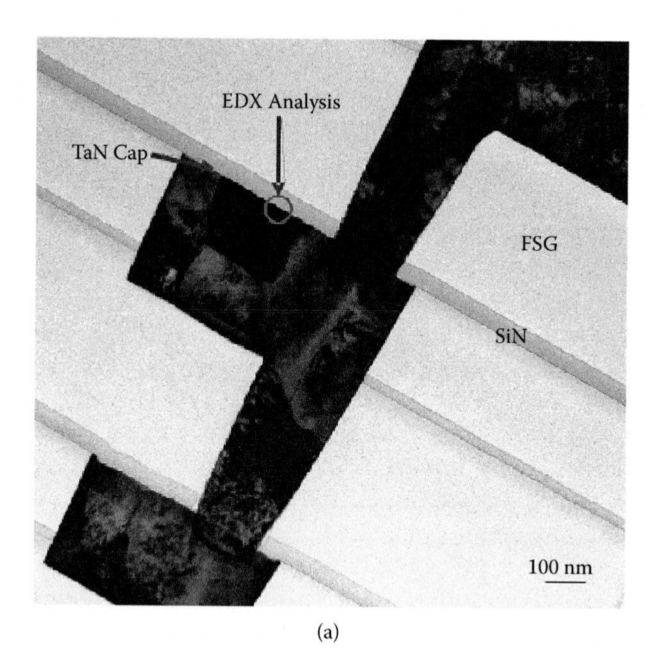

(a)

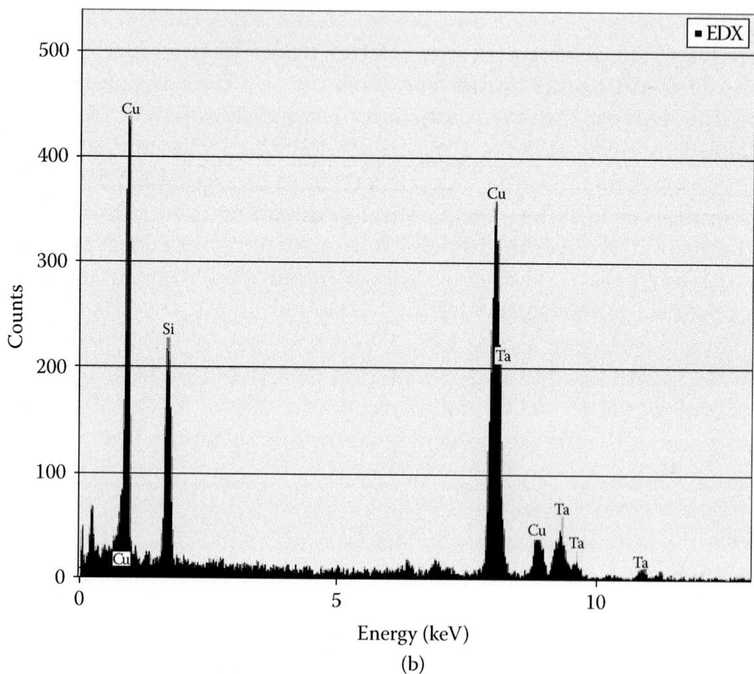

(b)

FIGURE 4.30 (a) The cross-sectional view of the Cu surface capped with a very thin layer of TaN. (b) The EDX analysis of the circled area of (a).

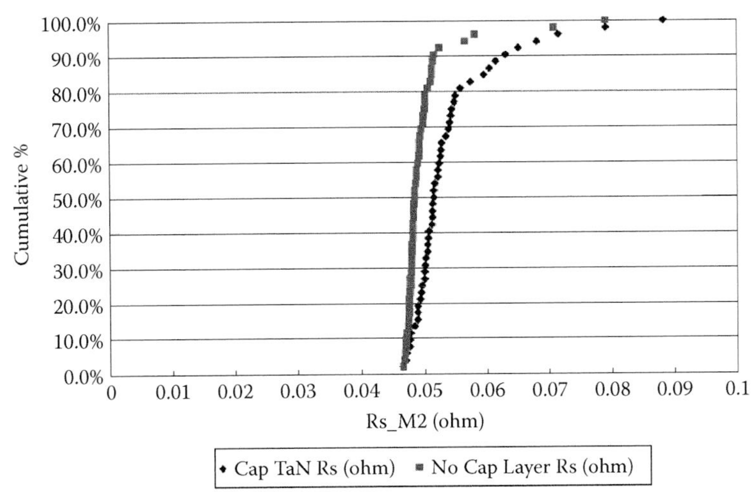

FIGURE 4.31 The Rs of specimens with TaN and without TaN capping at 0.19 μm in width and 1400 μm in length.

above the Cu and dielectric was polished away by a TaN module and a cross-sectional view of such a composite structure is given in Figure 4.30a. The analysis of a capping layer using electron dispersive spectroscopy (EDX) is shown in Figure 4.30b. After TaN capping, Cu was isolated from oxidative ambient and its corrosion was thereby prevented. However, the TaN capping raised the resistance of Cu (*Rs*). Because the composite structure contains a high resistivity TaN layer (100 times higher than the Cu), the *Rs* of the Cu-containing TaN cap provides a 7.9% increment on resistance than that without TaN cap, as shown in Figure 4.31. On the other hand, the uniformity of *Rs* of composite metal is also lower than that without TaN capping. Furthermore, the Cu CMP repolish would worsen the non-uniformity of *Rs*. It is well known that Cu reacts easier with oxygen. The oxidation is a continuous reaction due to porous nature of Cu oxide and raises the resistivity of the Cu. The more Cu oxidation occurred, the less speed gain from the material changing from Al to Cu. In addition, the process reliability as well as the lifetime of products will shorten. The corrosion defect characterizations shown in Figure 4.32 and Figure 4.33 reveal that the corrosion defects dramatically increase in the specimens without TaN capping. As for the specimens capped with TaN, the defect level remained the same up to 128-h heating treatment at 180°C. This observation evidenced that TaN capping could effectively isolate the Cu to prevent corrosion in an ambient environment. Thermal stability is another important issue for the utilization of Cu interconnection. Thermal stability of Cu was evaluated by the via resistance shift and the result is shown in Figure 4.34. After baking for 170 h in a furnace, the specimens capped with TaN exhibited a better thermal stability, as indicated by the *Rc* shift percentage characterization. During the following thermal process, grain growth of Cu occurred and

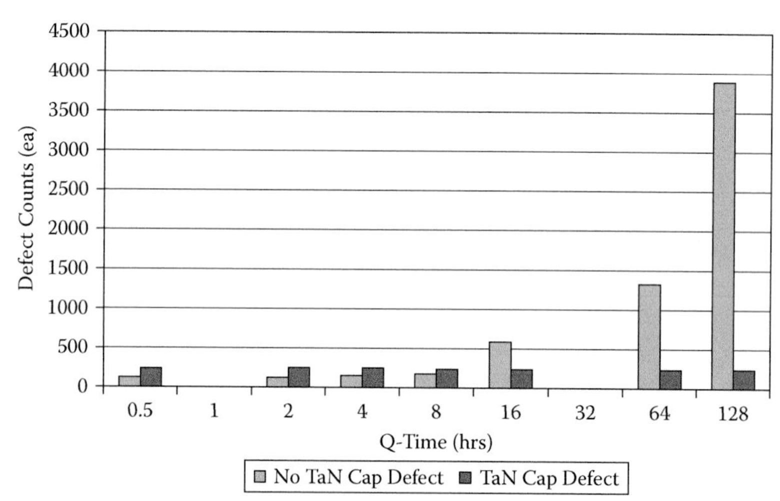

FIGURE 4.32 The relationship between waiting time and defect counts scanned by KLA-AIT-II. The defect counts increase with the waiting time.

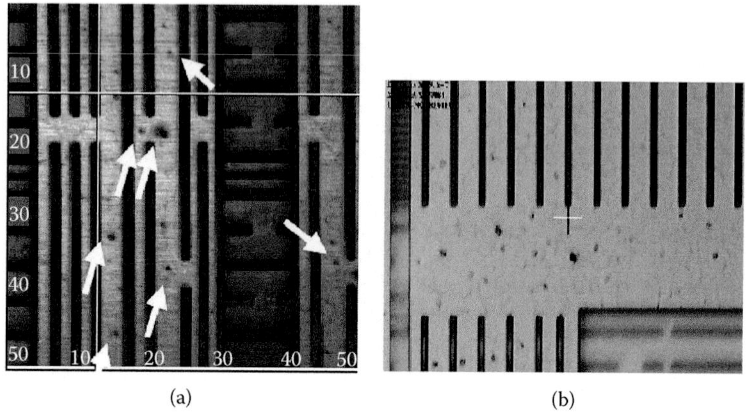

FIGURE 4.33 (a) The corrosion defects observed after annealing for 8 h at 180°C. (b) The corrosion defects observed after annealing for 128 h at 180°C.

the voids appeared. The interface of via connecting to prelayer metal is the preferential site of void formation. Poor adhesion between oxide and metal interface was observed at the bottom of via sidewall (i.e., the shrank), as shown in Figure 4.35. These voids deteriorated the thermal stability after high-temperature baking. The TaN capping is able to enhance the thermal stability because it restricts the Cu surface from reacting with the oxidative ambient and provides a good adhesion on the next Cu barrier layer, which is also of TaN. In addition, the TaN cap restricts the Cu line and inhibits its expansion during subsequent dielectric deposition. The restriction provides a stable volume of Cu during further thermal processes and hence leads to a higher thermal stability of Cu.

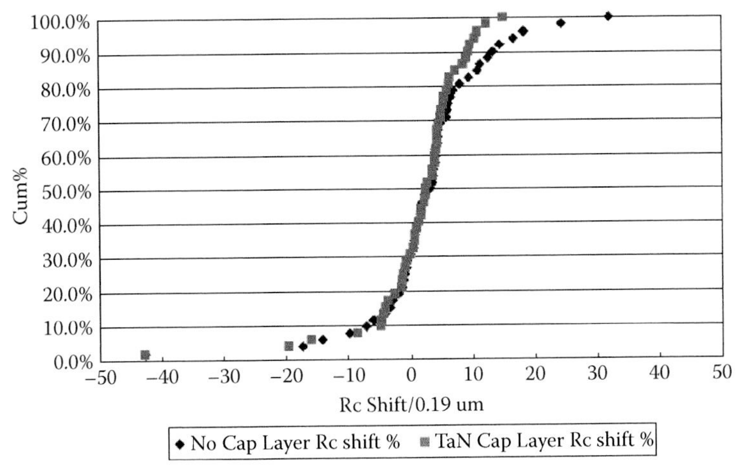

FIGURE 4.34 The shift of Rc after 180°C baking for 170 h.

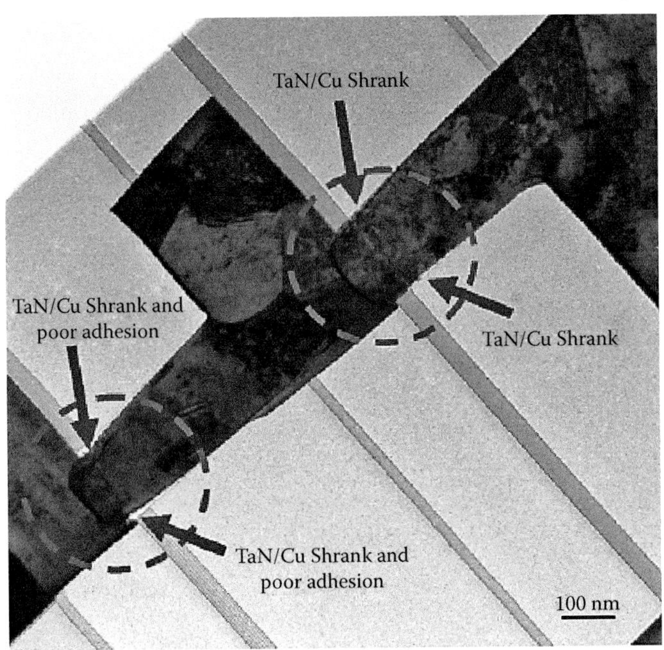

FIGURE 4.35 The via bottom (i.e., the shrank) exhibited a poor adhesion after heat treatment.

The amount of Cu dishing and erosion after Cu CMP was found to increase with the increases of Cu pad size and overpolish time. By capping a thin TaN layer on Cu, the Cu surface was effectively isolated from the oxidative ambient and the corrosion is presented. Furthermore, there is no increase of defect density in the specimens incorporating with the TaN capping process. The TaN capping process also provided a better thermal stability of Cu during subsequent thermal treatment. The only flaw of TaN capping is the increase of *Rs*, which will deteriorate the operation speed of devices. However, it could be overcome by design optimization of the circuits.

5

Nanotopography

5.1 What Is Nanotopography?

The industry definition of nanotopography covers a spatial wavelength range of approximately 0.2 to 20 mm. This region essentially defines surface topography dimensions larger than roughness. Roughness is defined as the smaller size spatial features on the wafer (an analogy to grains of sand in the desert), and nanotopology defines wider features up to approximately a lithographic field site. Nanotopography defines the nanometer-scale height variations that exist on lateral millimeter-scale wavelength on an unpatterned silicon wafer. The characteristics of the variation depend on the specific wafer manufacturing process used to generate a particular wafer. An illustration of wafer nanotopography is shown in Figure 5.1. The height variation is typically 100 nm with a lateral length scale between 1 and 20 mm. This is a parameter that measures the front-surface, free-state topology of an area that can range in size from fractions of a millimeter to tens of millimeters.

Nanotopography is the surface topography of wafers placed on a flat stage without chucking or clamping. It has been known as the waviness visually represented with an optical tool called a magic mirror, which provides images that are qualitative pictures of topography variations on the frontside surface of wafers. It has a peak-to-valley height that is considered to vary between several nanometers and several hundred nanometers, and its spatial wavelength range is considered approximately up to 20 mm.

In typical CMP machines, a front-side surface reference is employed, and the backside surface of a wafer touches soft carrier films or airbags, both of which absorb the topography variations of the backside surface. Thereby, wafers are in a condition without chucking or clamping in the CMP machines. Moreover, nanotopography differs from front-referenced site flatness in that for nanotopography the wafer is measured in a free state, while for flatness it is referenced to a flat chuck. A wafer may have perfect flatness (in the classical definition of flatness) yet still have nanotopography. If a wafer has surface irregularities on the front and backside

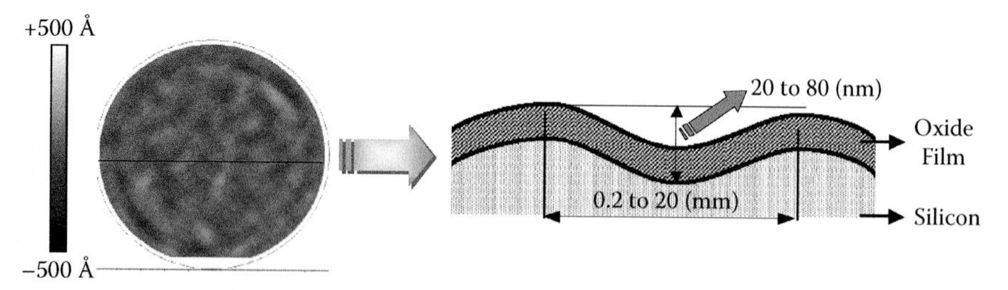

FIGURE 5.1 (See color insert) Illustration of wafer nanotopography.

of the wafer, but front and back surfaces are parallel, the wafer has perfect flatness. However, the same wafer will exhibit nanotopography (Figure 5.2).

Nanotopography bridges the gap between roughness and flatness in the topology map of wafer surface irregularities in spatial frequency (Figure 5.3). As linewidth shrinks with non-uniform pattern density and with the use of hard pads for CMP, nanotopography may significantly degrade the dielectric film uniformity.

Nanotopography is measured by two techniques: light scattering and interferometry. Light scattering tools typically employed for particle and surface-defect characterization can be used to measure the local slope change over the entire surface of the wafer. The local slope change may be integrated to yield height or topography information. Since the beam size can be on the order of fractions of a micron, nanotopography can be measured. Optical interference measurement is straightforward: A beam is split into two components—one component is reflected from the wafer surface and the second is reflected from a reference mirror. The interference of the combination of the two beams is a measurement of the topology of the wafer surface. With both techniques, signal filtering is used to separate the low-wavelength features (i.e., warp) so that only the

Normal Wafer Surface			
Backside Reference			
Flatness Nanotopography	Good Poor	Good Good	Poor Poor

FIGURE 5.2 Nanotopography variations on the frontside and backside condition.

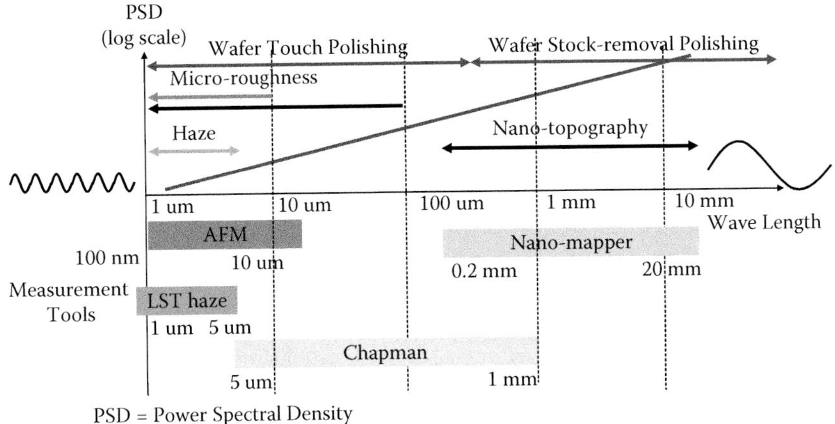

FIGURE 5.3 (See color insert) Topology map of wafer surface.

high-wavelength/low-frequency information, (i.e., the true surface nanotopography) is measured. The equipment used in measuring the nanotopography of the wafer will be introduced in Section 5.4.

5.2 Why Nanotopography Is Important

Recently, the nanotopography of the surface of silicon wafers has become an important issue because it may seriously affect the post-CMP uniformity of thickness variation of dielectrics.

Semiconductor device fabrication on silicon wafers comprises steps at which layers are deposited, subsequently planarized, and structured. Planarization is typically performed using a polishing step where the smoothing of the layer is due to chemical interaction with the polishing slurry as well as mechanical abrasion. Therefore, such processes are called chemical-mechanical polishing (CMP) and they have been implemented, for example, in silicon wafer manufacturing for more than 30 years. The homogeneity of a post-CMP layer is limited by fluctuations of the combination of layer deposition and CMP, as well as the frontside topography of the substrate. These two contributions to post-CMP layer thickness deviations of oxide film need to be quantified properly for identifying potentials that allow improvement of the efficiency of planarization for future devices. This is particularly necessary since excessively large-layer thickness variations after CMP may have a negative impact on device performance such as leakage and the pinhole effect, and the thickness of layers might even decrease with the ongoing reduction of critical dimensions.

Concern over local wafer site flatness for advanced lithography has led to new wafer surface topography requirements. These requirements are typically driven by CMP where film thickness variations can result in uneven surface topography. Control over local wafer site flatness becomes important for device geometries smaller than 0.25 μm.

5.3 Impact of Nanotopography on CMP

5.3.1 General Introduction

The interaction of nanotopography upon film polishing uniformity in CMP has been under extensive investigation by Boning and co-workers and Tamura et al. (2000). The primary effect of oxide uniformity removal is due to the hardness of the CMP pad. The fundamental concept is very simple: soft polishing pads conform to local topology variations (i.e., nanotopography), whereas hard pads do not. Figure 5.4, adapted from Boning et al. (2000), illustrates this principle. Typically, a wafer has a characteristic nanotopography length (NL, shown in the top illustration of Figure 5.4). The soft pad will conform over the nanotopography and maintain a uniform film. The hard pad will not conform to the nanotopography and produce a non-uniform film with high spots on the wafer surface having a thinner film and low spots having a thicker film. Traditionally soft pads

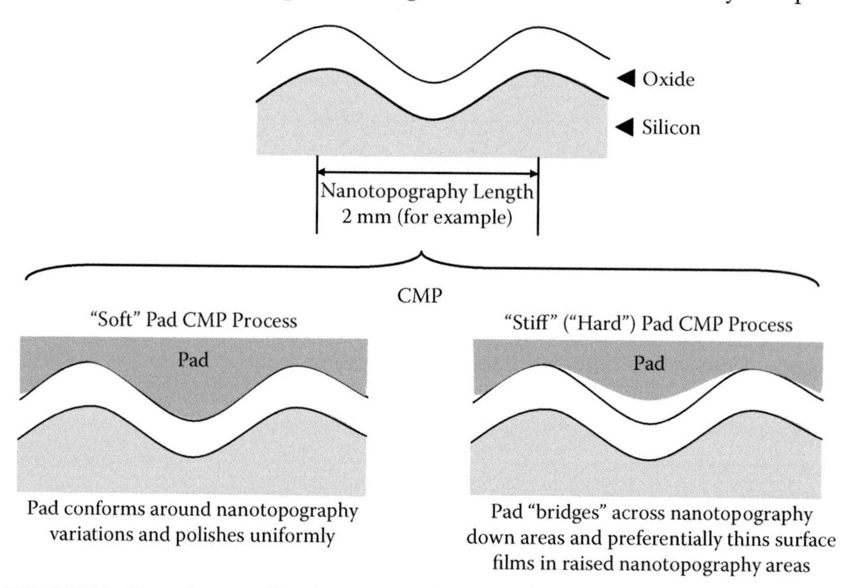

FIGURE 5.4 (See color insert) Basic concepts of soft and hard polishing pads.

have been used for film polishing in CMP. However, with the need for better planarization because of more layers, smaller critical dimension, and for multifunction logic devices that have several different areas of varying pattern densities, stiff pads are required. To some extent, the effect of nanotopography can be minimized by using polishing additives, such as ceria particles. Nonetheless, wafer nanotopography becomes increasingly important.

To understand the influence of nanotopography in CMP film polishing uniformity, the concept of planarization length should be considered. The planarization length (PL; Figure 5.5) is the spatial length at which polishing cannot reduce the step height of a feature in the film thickness, as shown in Figure 5.5. The important aspect to consider is when PL is less than NL, the film uniformity is maintained; however, when PL is more than NL one could find non-uniform film polishing. Two typical examples are shown in Figure 5.4. The CMP process and the film uniformity specifications may be considered to determine the level of nanotopography required.

Nanotopography of the silicon wafer is dictated to a large extent by the polishing process. For single-sided polished (SSP) wafers, the polishing process has been optimized to minimize nanotopography. In this process, to achieve good flatness, the wafer must be mounted or chucked against a flat reference block. Since the wafer backside is etched (not polished smooth), it has surface topology. Because of the fixing process used to mount the wafers (e.g., wax mounting or vacuum chucking), the topology of the backside of the wafer and the fixing surface or adhesive/wax are transmitted to the front side and causes nanotopography. The other technique of mounting a wafer (the one that is normally used in CMP), namely, free mounting, does not cause nanotopography formation, but also does not guarantee the wafer is made flat. The best flatness and nanotopography is obtained when the wafers are double-sided polished (DSP). The true, planetary, free-floating DSP process polishes both sides of a silicon wafer simultaneously. Since the wafer is polished in a free state,

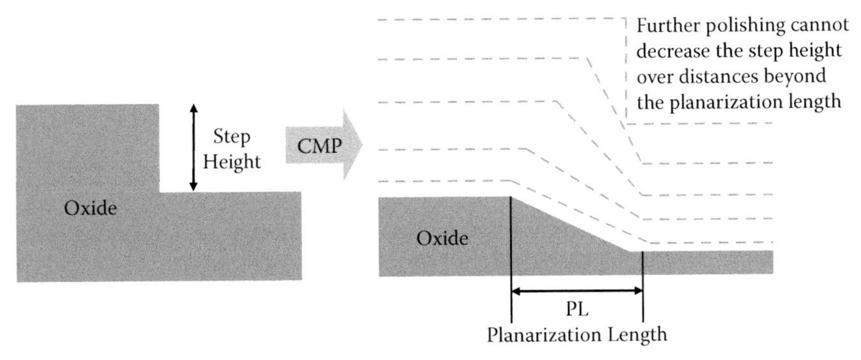

FIGURE 5.5 Basic concepts of soft (left) and hard (right) polishing pads.

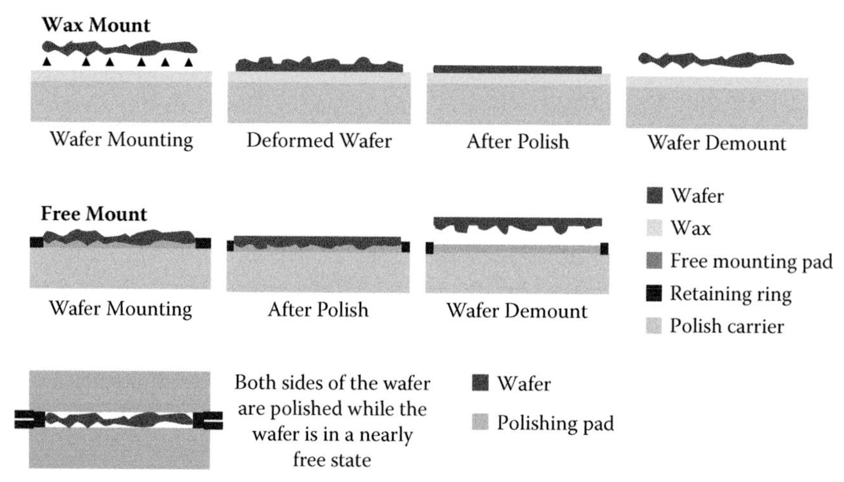

FIGURE 5.6 (See color insert) A comparison of SSP and DSP mounting techniques and how these affect nanotopography and flatness.

nanotopography is minimized. Also, good flatness is achieved. Thus, both good flatness and nanotopography are produced. Figure 5.6 shows a comparison of SSP and DSP mounting techniques and how these affect nanotopography and flatness.

Wafer manufacturing developed a planetary DSP process that provides wafers with leading-edge nanotopography and flatness characteristics that meet all of the increasing demands of CMP requirements. Although planetary DSP is technically the best method to achieve superior nanotopography and flatness, there are several barriers to practically applying this method in the fab. These barriers include cost of ownership for DSP, issues related to running both a polished backside DSP and an etched backside wafer in their lines at the same time, electrostatic chuck problems, and in-line sensor calibration. Wafer manufacturing companies are actively exploring both planetary DSP and SSP methods that promise to achieve a good balance between nanotopography and flatness results and cost of ownership.

5.3.2 Spectral Analysis of the Impact of Nanotopography on Oxide CMP and Fourier Transform Method

In general, it is quite convenient to apply spectral analysis to research the impact of nanotopography on film thickness variation in CMP. The surface height changed as the nanotopography of wafers were measured by an ADE NanoMapper. Figure 5.7 shows an example of nanotopography of a wafer measured by ADE NanoMapper. The darker region in the map corresponds to a lower height and the brighter region to a higher height.

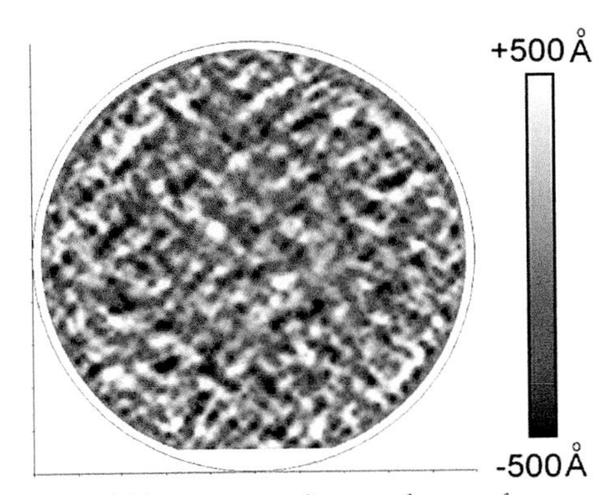

FIGURE 5.7 (See color insert) Nanotopography map for a wafer.

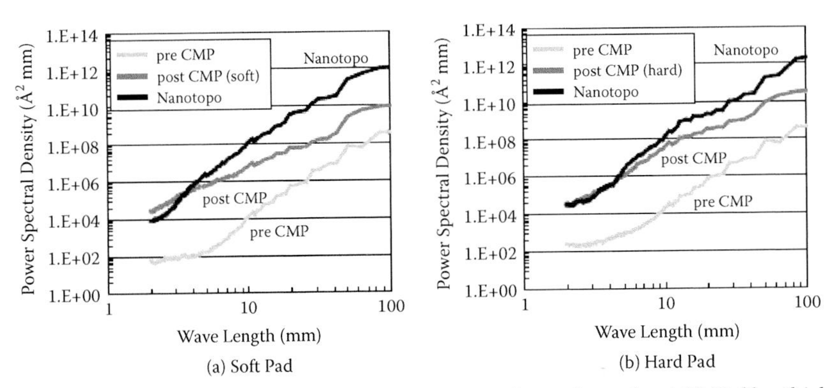

FIGURE 5.8 Power spectral densities of nanotopography and pre-/post-CMP film thickness variation with different pads.

Figure 5.8 shows the power spectral densities (PSDs) of nanotopography and pre-/post-CMP film thickness variation. The higher the PSD number, the higher nanotopography in the surface of the wafer. To directly relate the nanotopography and the film thickness variation after CMP, we introduce a theory conducted by the professor J.G. Park, who used Fourier transform function to convert PSDs to a more understandable parameter called transfer function $T(\mu, t)$.

Park and co-workers (2001) proposed a formula that describes the relationship between the nanotopography and the film thickness variations after CMP. The discussion is based on the concept of the planarization length of CMP, which is mentioned in Section of 5.3.1. Park et al. only deal with blanket (unpatterned, one material) wafers and the uniformity of the local polishing rate causing the fluctuations of surface height. The following formula is for one dimension:

$$h(x,t) = w(x) + f(x,t) \tag{5.1}$$

The height of the oxide surface at any time during polishing can be denoted as: x is position, t is time, $h(x, t)$ is the height of the oxide surface, $w(x)$ is the nanotopography of the silicon surface, and $f(x, t)$ is the film thickness of the oxide. To simplify the formulation, the origin of the height axes for $h(x, t)$, $w(x)$, and $f(x, t)$ can be arbitrary changed.

The raised regions on the surface may cause the dynamical excess deformation of the polishing pad incurred during polishing. The deformation is propagated within a certain lateral range and it is a nonlocal phenomenon. It can be described by the spatial-convolution function in the real domain. The excess deformation of the polishing pad may affect the local removal rate. Supposing its effect is the linear correlation, the local polishing rate can be given by:

$$\frac{\partial h(x,t)}{\partial t} = -\left(K_0 + K_1 \cdot CMP_1 \otimes h(x,t)\right) \tag{5.2}$$

where K_0 is the blanket polishing rate without considering nanotopography, K_1 is the coefficient for the local excess-polishing rate caused by surface-height fluctuation, CMP_1 (r) is the response function (r is the lateral distance from the origin) describing the nonlocal pad deformation and its affect on the polishing rate. The function CMP_1 (r) is an even function and closely related to polishing parameters, such as the polishing-pad hardness, relative velocity between pad and wafer, polishing pressure, and so forth, through the planarization length that characterizes the lateral range over which the raised topography interacts. After integration of Equation 5.2 for time, $h(x, t)$ is expressed as:

$$h(x,t) = h(x,0) - \left\{K_0 \cdot t + K_1 \int_0^t CMP_1 \otimes h(x,t)dt\right\} \tag{5.3}$$

The unknown function $h(x, t)$ is too complicated to solve analytically, so a simplification is applied below.

When $t = 0$, if the film thickness $f(x,0)$ is uniform enough for our analysis, the initial surface-height fluctuation can be denoted as follows using Equation 5.1 (Note: the origin of the height axis is variable).

$$h(x,0) = w(x) \tag{5.4}$$

The spatial profile of $h(x, t)$ is continuously changing as the polishing is performed. However, supposing that the $h(x, t)$ can be described as the convolution of $h(x, 0)$ (as the initial profile) and another time-dependent proper response function CMP_2 (r, t), the equation is given by:

$$CMP_1 \otimes h(x,t) = CMP_2(t) \otimes w(x) \tag{5.5}$$

Then, Equation 5.3 can be rewritten as:

$$\begin{aligned} h(x,t) &= w(x) - K_1 \int_0^t CMP_2(t) \otimes w(x)dt \\ &= w(x) - \left(K_1 \int_0^t CMP_2(t)dt) \right) \otimes w(x) \\ &\equiv w(x) - CMP(t) \otimes w(x) \end{aligned} \tag{5.6}$$

where *CMP* (*r*, *t*) is a time-dependent spatial-convolution function, and the term of K_0 in Equation 5.3 is omitted because it is *x*-independent. After Equations 5.1 and 5.6, the film thickness *f*(*x*, *t*) can be expressed as:

$$f(x,t) = -CMP(t) \otimes w(x) \tag{5.7}$$

Next, the Fourier transformation of Equation 5.7 is denoted as:

$$F(\lambda,t) = -FTCMP(\lambda,t) \cdot W(\lambda) \tag{5.8}$$

where λ is the spatial wavelength, F(λ, *t*), *W*(λ), and *FTCMP*(*t*) are the Fourier transforms of *f*(*x*, *t*), *w*(*x*), and *CMP*(*t*), respectively. After Equation 5.8, the relationship between the power spectral densities of film thickness variation and the wafer nanotopography can be expressed as:

$$|F(\lambda,t)|^2 = T(\lambda,t) \cdot |W(\lambda)|^2 \tag{5.9}$$

where T(λ, t) is $|\mathrm{FTCMP}(\lambda, t)|^2$, and it is a time-dependent response function and should still have a characteristics length originating the planarization length, which depends on the CMP parameters mentioned earlier. This response function T(λ, t) can be regarded as the transfer function from the nanotopography spectrum to the film thickness variation spectrum. That is to say, if this transfer function has a large value in a certain wavelength region, the component of the nanotopography impacts severely on the film thickness variation after CMP.

Figure 5.9 is an example of two calculated transfer functions with wavelength. As mentioned earlier, the larger value of transfer function in certain wavelengths means the stronger impact of nanotopography on film thickness variation.

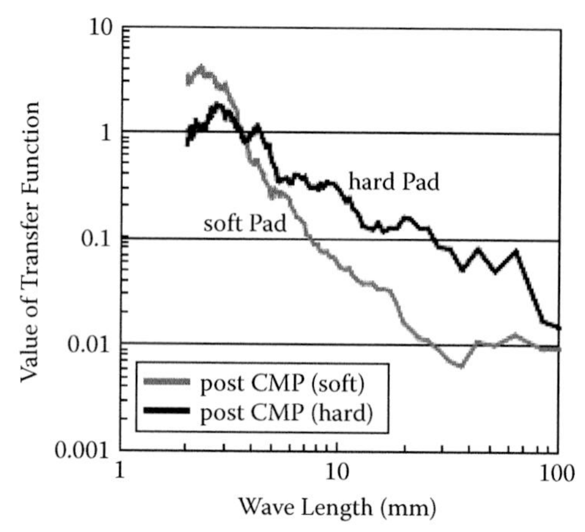

FIGURE 5.9 Transfer function as the ratio of the oxide film thickness variation to that of nanotopography.

The more detailed analysis of the Fourier transform function of Equation 5.9 and the correlation between T(λ, t) and the nanotopography impact on the film thickness variation with different pad hardness will be discussed in the next section.

5.3.3 Impact of Nanotopography on Silicon Wafer on Oxide CMP

5.3.3.1 Wafering Method Dependency of Impact of Nanotopography on Oxide CMP

In addition to CMP process optimization, better silicon wafers are required to solve the nanotopography problem. In general wafer manufacturing (wafering), chemical etching is applied to remove the mechanical damage induced during the preceding lapping process. This is followed by a polishing process to improve the parallelism of the wafer and create a very flat surface. The polishing process applied during wafering can be either DSP or SSP. For 8-inch wafers, device manufacturers generally require SSP because the frontside of the wafer is then distinct from the backside. To improve the nanotopography of SSP wafers, a combined chemical etching and SSP process that can be applied during wafering was developed, and the performance of the resulting wafers for oxide CMP was examined by the analysis method described earlier.

Three chemical etching methods were applied: acid etching, alkali etching, and multietching (acid and alkali), which allowed us to obtain the benefits of both acid and alkali etching in a compatible process. To prepare the wafers to be tested, we used two polishing methods: conventional single-side polishing with wax mounting (SSP1) and improved single-side polishing (SSP2), in which the backside topography has little effect on the polishing of the front surface. Combining the three etching methods and the two polishing methods yielded six types of sample wafers. In this book, we will use a notation based on the letters and numbers given above to distinguish between wafer types, that is, A1 refers to a wafer that underwent acid chemical etching and wax mounted polishing (SSP1).

Professor J.G. Park has produced a series of experiments to understand the effect of the different etching method and the different polishing method on film thickness variation after the CMP process. Figure 5.10 shows the height maps and line profiles of the six wafer types. Table 5.1 shows the parameters in this experiment of oxide CMP.

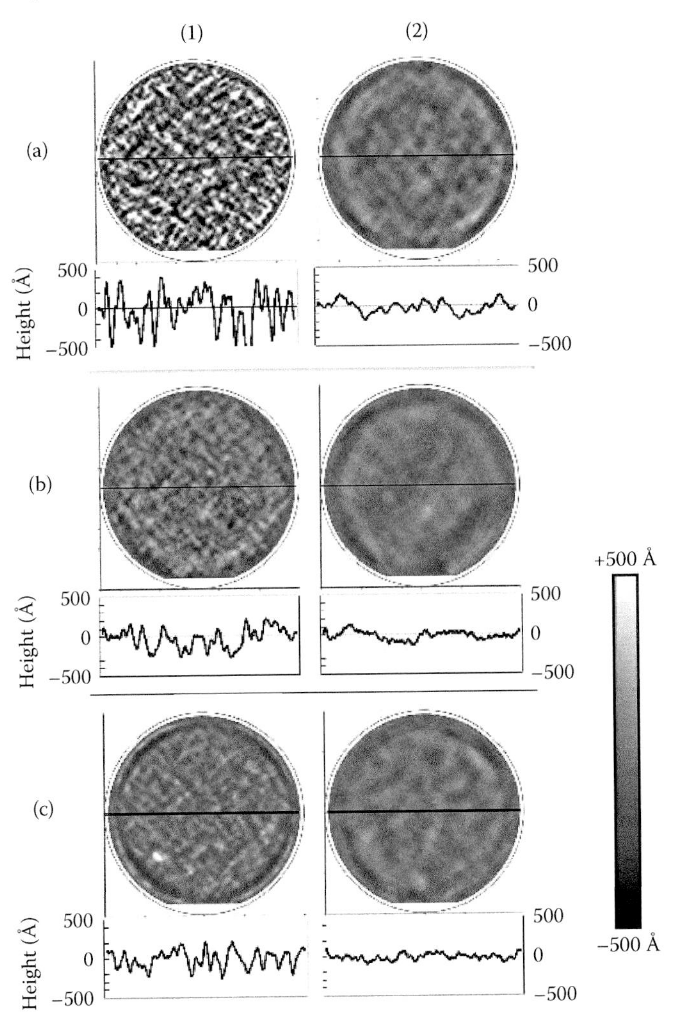

FIGURE 5.10 Nanotopography maps of wafers prepared using each process combination. Darker regions of the map correspond to a lower height and brighter regions to a greater height. The height profiles along the x-axis are shown.

TABLE 5.1
Standard Deviation of Height Profile (nm)

	SQM and NanoMapper	NanoMapper and DynaSearch	DynaSearch and SQM
Measurement 1	23.29	17.20	29.57
Measurement 2	13.00	13.21	18.23
Measurement 3	17.85	16.53	26.55
Measurement 4	12.15	11.57	19.95
Measurement 5	14.40	14.62	21.20

As shown in Figure 5.10, the A1 wafer had the largest nanotopography, whereas the C2 wafer had the smallest. Comparing the polishing methods, we observe that SSP2 resulted in smaller height variation than SSP1. Figure 5.11 shows examples of the correlation between nanotopographic

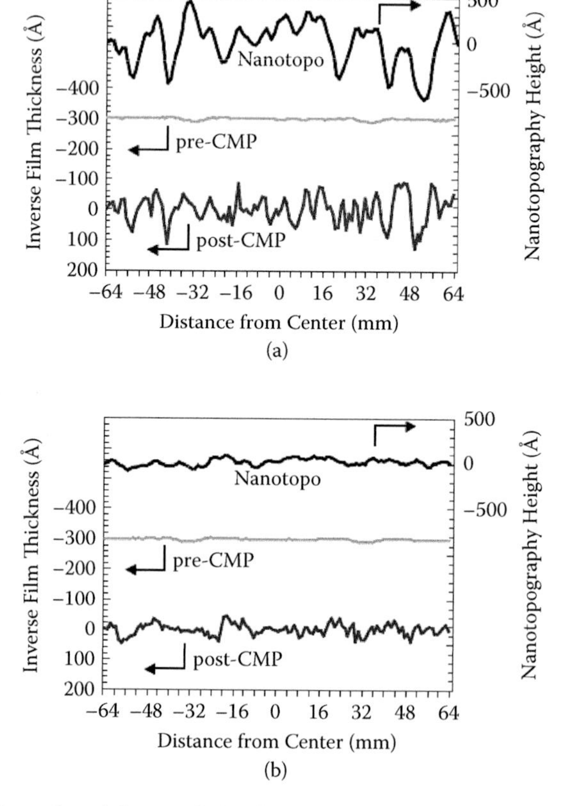

FIGURE 5.11 Examples of the correlation between the nanotopography and inverse film thickness profiles before and after CMP for (a) A1 wafer and (b) C2 wafer.

line profiles and film thickness variation before and after CMP. Figure 5.11a shows that for the A1 wafer and Figure 5.11b shows the C2 wafer. The peak and valley positions of the nanotopography coincided well with the film thickness variation after CMP, particularly for the A1 wafer. It can be observed that even the small nanotopography of the C2 wafer slightly influenced the film thickness variation. We attributed the fluctuations in the film thickness after CMP to the nanotopography. Note that the film thickness profiles are inversely plotted, which means that excess thinning occurred due to a pressure concentration at each local peak position of the nanotopography.

In Figure 5.12, the standard deviation of the nanotopography and the film thickness variation after CMP are summarized for the six wafer types. Each value was calculated from the filtered profiles. For all types of etching, particularly acid etching, SSP2 more effectively reduced the nanotopography impact on the post-CMP film thickness variation. Wax mounting was

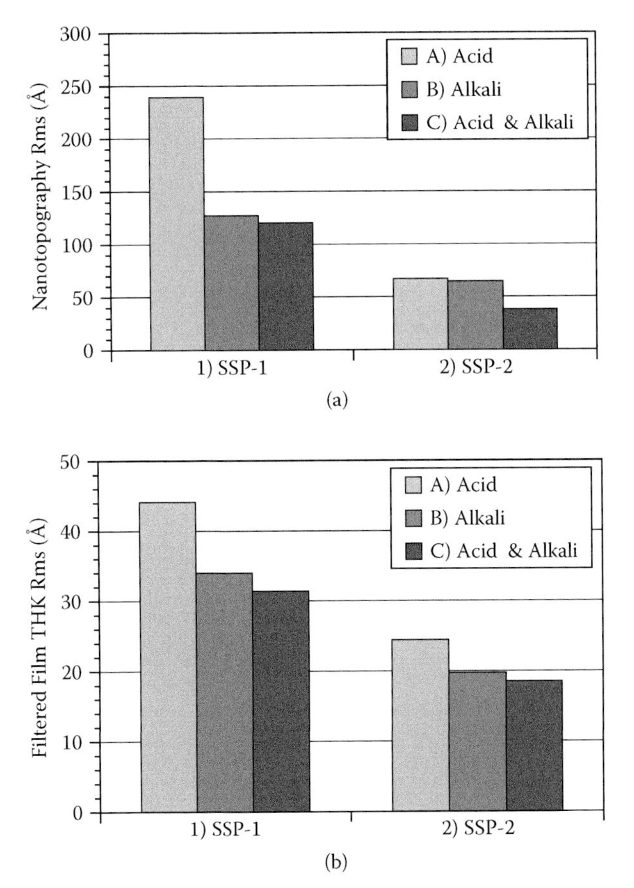

FIGURE 5.12 Standard deviations (Rms) of (a) the nanotopography and (b) the filtered film thickness after CMP.

used in SSP1 to hold the backside of the wafer on the flat ceramic plates. The backside waviness due to the chemical etching caused excessive height variation on the front surface through the internal stress distribution within the wafer. After stock removal polishing, the front surface became smooth, but once the wafer was removed from the ceramic plate, the internal stress was released and variation of the front surface height caused by the backside waviness appeared. This is probably the reason for the SSP1 wafers showing more nanotopographic variation than the SSP2 wafers. Among the different etching methods, acid etching resulted in the greatest nanotopographic variation, whereas the combined acid and alkali etching would account for the greater nanotopographic variation. Multietching can be optimized to reduce the waviness by balancing the acid and alkali factors. The variation that appears as waviness after the chemical etching leads to nanotopography due to the backside influence described above and/or residual waviness on the front side after polishing.

The correlation between the standard deviations of the OTD (oxide thickness deviation) and the NH (nanotopography height) is plotted in Figure 5.13. The OTD is independent of the nanotopography before CMP. On the other hand, the OTD and NH after CMP show a clear, positive correlation. The slope of the fitting line for the post-CMP correlation in Figure 5.13 may vary, depending on the CMP process parameters. A small slope indicates that the nanotopography is less likely to have an impact on the effectiveness of the CMP process.

We also calculated the PSDs of the NH and OTD. The PSDs for the A1 and C2 wafers are shown in Figure 5.14 as examples. This analysis quantitatively demonstrates that the C2 (multietching, SSP2) wafer had less nanotopography than the A1 wafer within the wavelength range from 2 to 100

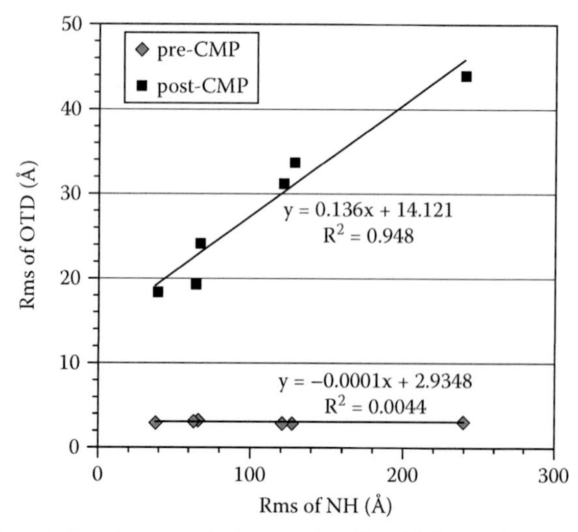

FIGURE 5.13 Correlations between the Rms of the NH and that of the OTD.

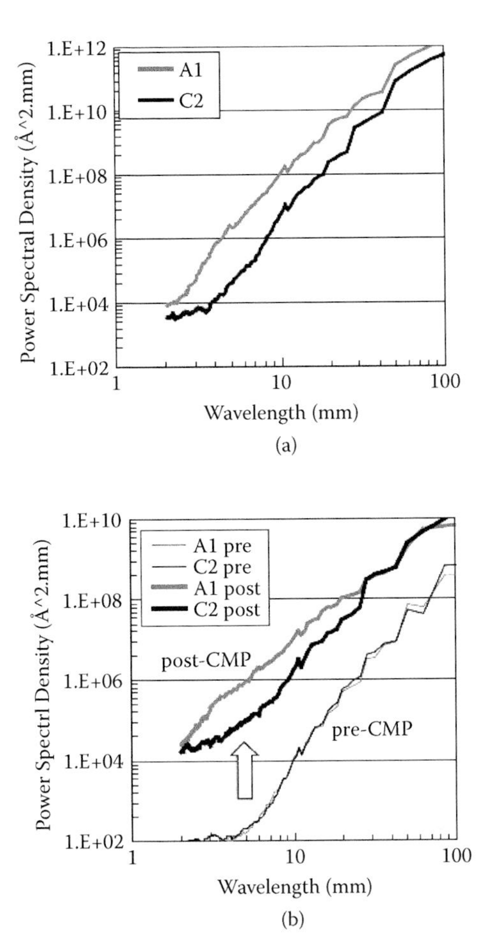

FIGURE 5.14 Power spectral densities of (a) height changes and (b) film thickness variations before and after CMP for the A1 and C2 wafers.

mm (Figure 5.12a). Regarding the film thickness variation (Figure 5.12b), the spectra before CMP were independent of the wafer nanotopography. On the other hand, the PSD of the OTD increased after CMP in accordance with the wafer nanotopography, which clearly demonstrates that the C2 wafer had a more uniform film thickness than the A1 wafer within the wavelength range from 2 to 30 mm.

A clear and positive correlation between the standard deviation of the nanotopographic profile and that of the film thickness after CMP is evident. The PSD analysis of the CMP results quantitatively demonstrated the effect of the nanotopography of different wafers on the film thickness variation. Wafers prepared with the improved SSP technique had less film thickness variation after unpatterned oxide CMP than the wafers that underwent conventional SSP due to the impact of the nanotopography.

5.3.3.2 Slurry Characteristic Dependency of Impact of Nanotopography on Oxide CMP

To quantitatively analyze the impact of nanotopography on post-CMP oxide film thickness, we have introduced a spatial spectral method, and have used it to examine the effect of the pad type, removal depth, wafer manufacturing technique, and polishing method. However, the role of the slurry in controlling the impact of nanotopography on STI CMP is not yet clear. In this section, we discuss how the concentration of surfactant and abrasive size affect the impact of nanotopography.

Four kinds of slurry with different sizes of abrasives, denoted as A, B, C, and D, were prepared through a mechanical treatment. Figure 5.15 shows the correlation between the lateral profiles of the NH and post-CMP OTD on a wafer for two surfactant concentrations. Figures 5.15a and b correspond to slurry A (largest abrasives) with surfactant concentrations of 0 and 0.80 wt%, respectively. Figure 5.15c and d correspond to slurry D (smallest abrasives), also with surfactant concentrations of 0 and 0.80 wt%, respectively. For each of the various surfactant concentrations, the peak and valley positions of the NH and post-CMP OTD coincide well with each other. Therefore, the fluctuations in the post-CMP OTD can be attributed to the wafer nanotopography. Whereas the magnitude of the OTD for slurry A is similar with or without surfactant, that for slurry D increased with the surfactant concentration.

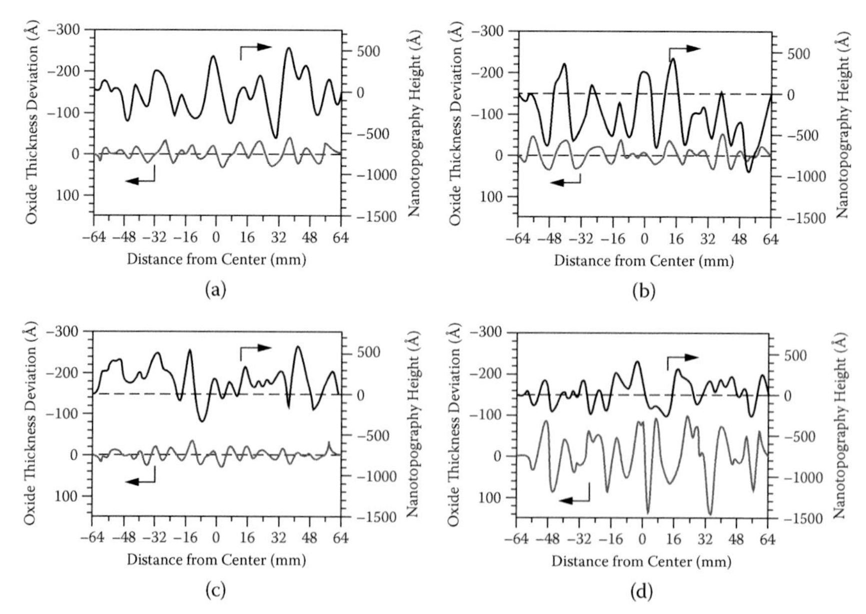

FIGURE 5.15 Correlation between the NH and the post-CMP oxide thickness variation: (a) Slurry A, 0 wt%; (b) Slurry A, 0.8 wt%; (c) Slurry B, 0 wt%; (d) Slurry B, 0.8 wt%.

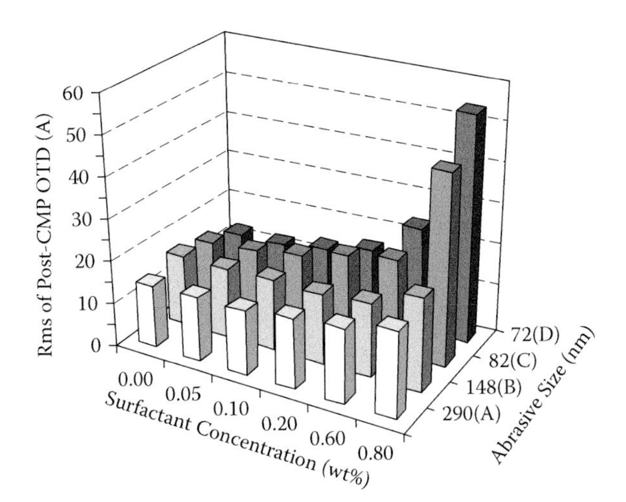

FIGURE 5.16 Standard deviation (Rms) of the post-CMP oxide thickness deviation.

The correlation between the abrasive size, the surfactant concentration, and the standard deviation of the OTD is shown in Figure 5.16. For slurries A and B, the standard deviation of the OTD is not influenced by the surfactant concentration. For slurry C, however, the standard deviation increases gradually with increasing surfactant concentration. For slurry D, it increases even more strongly with increasing surfactant concentration. Thus, when slurry with smaller abrasives is used, the magnitude of the OTD increases with increasing surfactant concentration. That is, the surfactant more strongly influences the nanotopography impact on OTD after CMP in the case of smaller abrasives. Though other factors, such as the selectivity of the removal rate between oxide and nitride films, must be taken into account when discussing the influence on the actual STI CMP process, these findings show that the nanotopography impact can be controlled by manipulating the slurry characteristics. Note that even if the ceria slurry generally used in STI CMP has high oxide-to-nitride removal selectivity, it basically avoids only excessive thinning of the nitride. It does not avoid incomplete clearing of the oxide as an influence on the nanotopography.

The PSDs of the post-CMP OTD for three surfactant concentrations are shown in Figure 5.17. The PSD increased with increases in the surfactant concentration in the wavelength range up to 30 mm. This result suggests that the change in planarization efficiency (or nanotopography impact) as a result of adding the surfactant occurs even in a longer wavelength range up to around 30 mm.

The mechanism for the impact of nanotopography on post-CMP OTD is directly related to the planarization of the oxide surface. That is, the local polishing rate of the protruding areas produced by nanotopography is greater than that of the valley areas, which causes excessive thinning of the oxide film at each nanotopography peak, as shown in Figure 5.18.

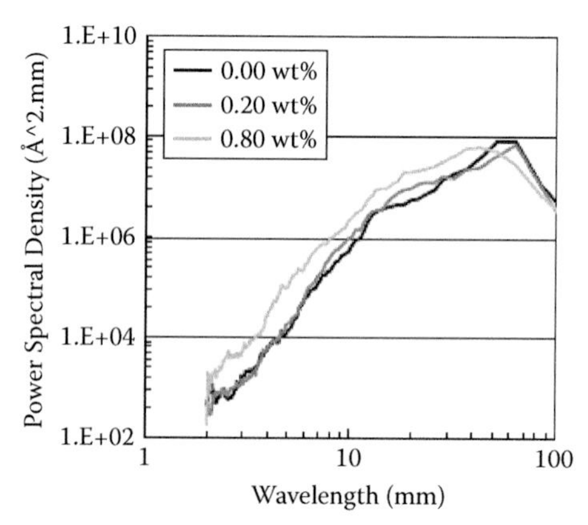

FIGURE 5.17 Power spectral densities of oxide thickness deviation after CMP.

Boning and Lee (2002) simulated this nanotopography impact through a contact mechanics model of the local contact pressure of the polishing pad on the wafer surface. However, an analysis that considers only the polishing pad and the film surface topography cannot explain the results of this study, as the role of the surfactant in planarization must also be taken into account. Nojo et al. (1996) proposed a useful model that contributes to our understanding of these results. They reported that adding a surfactant to ceria slurry can result in "self-stopping polishing," in which the polishing rate drops automatically as planarization progresses. They attributed this to the protective layer formed on the surface of the film by the surfactant. We expanded this model to take account of the effect of the abrasive size, as shown in Figure 5.18a. The larger abrasives marked A and B in the figure can remove both protruding areas and valley areas, but the smaller abrasives marked C and D can remove only protruding areas due to the thicker surfactant layer when the surfactant is added to the slurry. Accordingly, for the smaller abrasives case (marked C and D), the surface of the post-CMP film in the slurry with surfactant becomes flatter than that in the slurry without surfactant. A flatter post-CMP film surface corresponds to a more severe impact of nanotopography on OTD, as shown in Figure 5.18b. As a result, this model can explain the results shown in Figure 5.16.

In conclusion, the magnitude of film thickness variation after CMP was found to increase with the surfactant concentration in slurries with smaller abrasives but to be almost independent of the surfactant concentration in the slurry with the larger abrasive. This result can be explained with the model based on the passivation layer of the surfactant adsorbed on the oxide film surfaces during polishing.

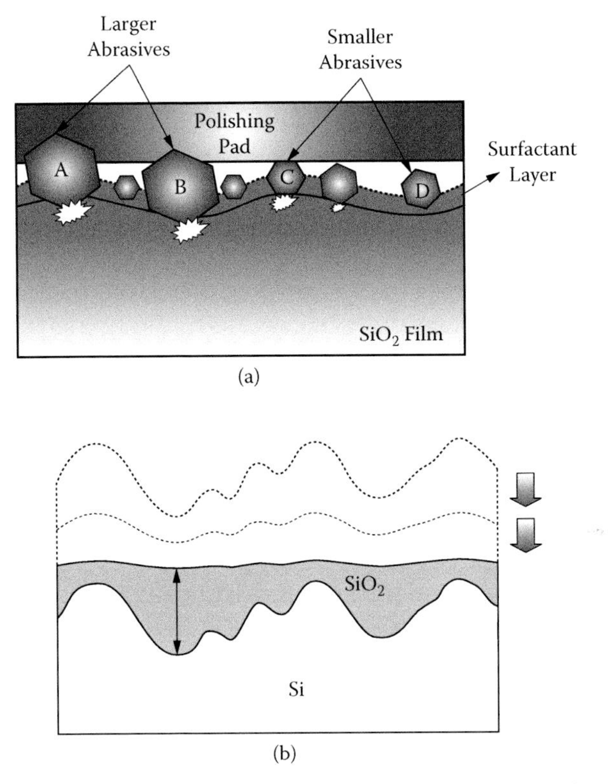

FIGURE 5.18 Proposed model for the nanotopography impact: (a) dependence of surface planarization efficiency on the abrasive size and surfactant and (b) relationship between film surface flatness and oxide thickness deviation after CMP.

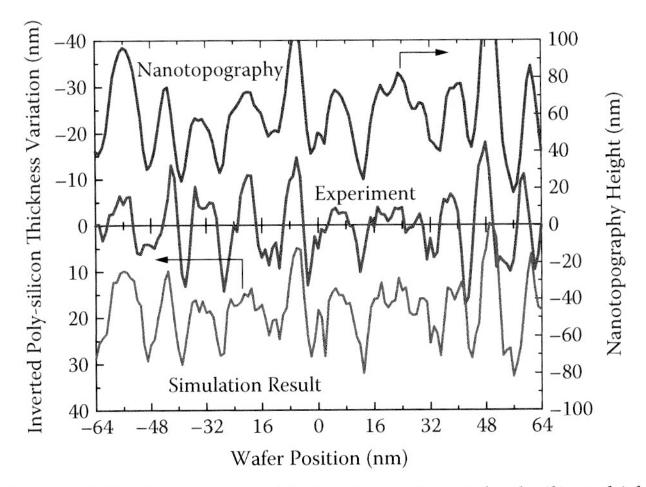

FIGURE 5.19 Correlation between simulated and experimental polysilicon thickness variation reduced by wafer nanotopography.

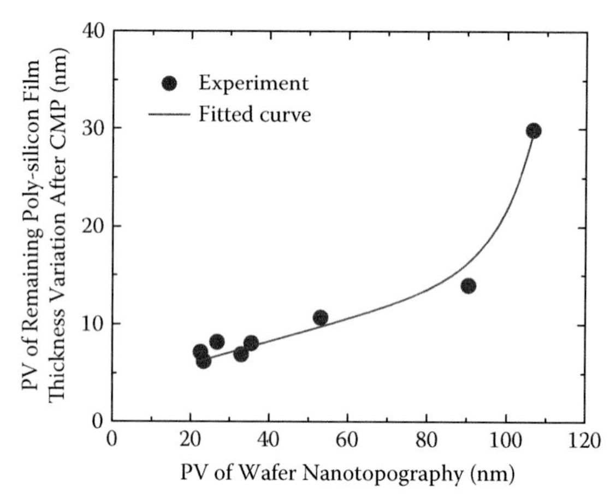

FIGURE 5.20 PV of remaining polysilicon thickness after polysilicon CMP, depending on PV of wafer nanotopography in 10-mm scan.

5.3.3.3 Effect of Wafer Nanotopography on Remaining Polysilicon Thickness Variation after Polysilicon CMP

Figure 5.19 shows the centerline profile for the wafer nanotopography (top line profile) and that for the remaining polysilicon thickness variation (middle profile line). The peak and valley positions of inverted thickness variation are well correlated with that of wafer nanotopography. This indicates that wafer nanotopography directly influences the thickness variation after polysilicon CMP. The bottom line in Figure 5.19 shows the simulated remaining polysilicon thickness variation induced by wafer nanotopography, based on the wear-contact model. It is obvious that the measured thickness variation (middle profile line) is well correlated with the simulated one (bottom line profile). This indicates that the polysilicon CMP mechanism using colloidal silica slurry follows Prestonian behavior rather than non-Prestonian behavior. Figure 5.20 shows the peak-to-valley (PV) value along the 10-mm radial scan of the remaining polysilicon thickness variation after polysilicon CMP as a function of the PV value along the 10-mm radial scan of wafer nanotopography. The PV value after polysilicon CMP exponentially increases with that of wafer nanotopography. Again, the wafer nanotopography strongly affects the remaining polysilicon thickness variation after polysilicon CMP, suggesting that the wafer nanotopography results in V_T variation for a NAND-flash memory-cell fabricated with the self-alignment of polysilicon floating-gate via polysilicon CMP. This is because both the voltage coupling of the floating-gate and the floating-gate interference in the NAND-flash memory-cell are determined by the height of the polysilicon floating-gate, which is influenced by the remaining polysilicon thickness variation after polysilicon CMP.

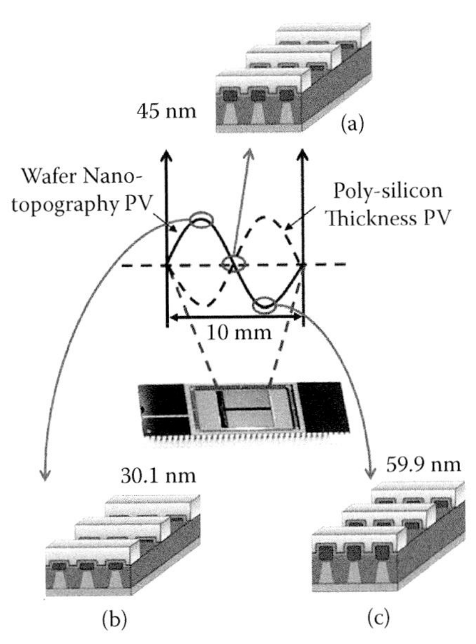

FIGURE 5.21 (See color insert) Schematic structures of a 63-nm NAND-flash memory-cell with different floating-gate heights, induced by the wafer nanotopography of 10-mm-diameter scanning: (a) 45-nm height (free of wafer nanotopography influence); (b) 30.1-nm height (at the top of wafer nanotopography influence); (c) 59.9-nm height (at the bottom of nanotopography influence).

5.3.3.4 Effect of V_T Variation of Wafer Nanotopography on Remaining Polysilicon Thickness Variation after Polysilicon CMP

A 63-nm NAND-flash memory-cell has a chip-side size of approximately 10 mm, as shown in Figure 5.21. We assumed that after polysilicon CMP the peak position of wafer nanotopography produces the smallest height of the polysilicon floating-gate (Figure 5.21b) and that the valley value of wafer nanotopography produced the largest height of polysilicon floating-gate (Figure 5.21c). To calculate the V_T variation originating from the PV value of the remaining polysilicon thickness after polysilicon CMP induced by wafer nanotopography, we should extract the floating-gate voltage determined by the floating-gate voltage coupling ratio and the floating-gate interference effect. Figures 5.22a and b show a three-dimensional schematic drawing of NAND-flash memory-cells and top views of each cell. Here, it was assumed that memory cell No. 5 was programmed. The applied voltage on the control gate, V_{CG}, at the selected word line was 18 V (V_{pgm}), while that at the unselected word line was 9 V (V_{pass}) for the 63-nm cell. The selected word-line bias was 16 V (V_{pgm}), and the unselected word-line bias was 8 V (V_{pass}). The applied voltage on the selected bit line was ground (GND), while that at the unselected bit line was 1.8 V (V_{cc}). Therefore, the channel-regions bias (V_{ch}) of unselected bit-line cells was floating.

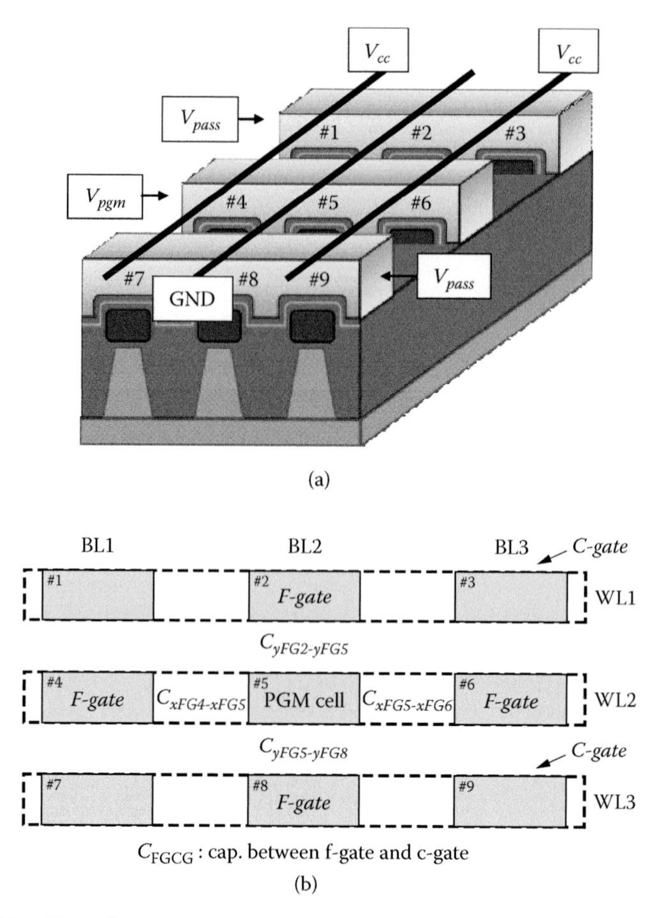

FIGURE 5.22 (See color insert) Programming operation of a NAND-flash memory-cell, where the programming cell is No. 5: (a) voltage bias conditions for memory cells and (b) parasitic capacitances during No. 5 cell programming.

The device dimension for the 45-nm cell was extracted from the dimensional trend for 90-, 70-, and 63-nm technology nodes. According to this trend, the height, length, and width of the floating-gate are 32, 45, and 57 nm, respectively. For the 63-nm NAND cell, oxide–nitride–oxide (ONO) layer and tunneling oxide thicknesses (equivalent oxide thickness, EOT) were 14.5 and 6 nm; for the 45-nm one, they were 11.9 and 5.2 nm, respectively. Each of the parasitic capacitances was obtained using the equation $C = \varepsilon_r\varepsilon_0 A/d$, where ε_r, ε_0, A, and d are the dielectric constant for insulating film, dielectric constant for vacuum, capacitor area, and insulation film thickness, respectively. $C_{xFG4\text{-}xFG5}$, $C_{xFG5\text{-}xFG6}$, $C_{yFG2\text{-}yFG5}$, and $C_{yFG5\text{-}yFG8}$ are the parasitic capacitances between the floating-gates at cells 4 and 5, 5 and 6, 2 and 5, and 5 and 8, respectively. $C_{yFG2\text{-}yCG5}$ and $C_{yFG5\text{-}yCG8}$ are the parasitic capacitance between the floating-gate at cell 2 and control-gate at cell 5 and between the floating-gate at cell 5 and control-gate at cell 8, respectively. The other parasitic capacitances were neglected because their values are compared with those of the above-mentioned six parasitic capacitances. Since the ONO capacitance at the programmed memory-cell (C_{ONO}) depending on the height of floating-gate is related to the voltage coupling ratio, the calculation of C_{ONO} was performed with a device simulator (ATLAS, Silvaco Corp.) Using the results of two-dimensional Poisson equation solving, C_{ONO} was calculated by the formula

$$V_{FG} = \frac{C_{ONO}}{C_{ONO} + C_{TUN}} V_{CG} \rightarrow C_{ONO} = \frac{C_{TUN} V_{FG}}{V_{CG} - V_{FG}} \tag{5.10}$$

where V_{FG} and C_{TUN} are the floating-gate voltage without considering the floating-gate interference effect and tunnel oxide capacitance. As shown in Figure 5.23, C_{ONO} for the 63- and 45-nm memory-cell linearly increases with increasing polysilicon floating-gate height determined by the PV of the remaining polysilicon thickness after polysilicon CMP. After the calculation of C_{ONO}, the floating-gate voltage, considering the floating-gate interference effect (V_{FG5}), was calculated by

$$V_{FG5} = \frac{C_{ONO} V_{CG5} + C_{xFG4-xFG5} V_{FG4} + C_{xFG5-xFG6} V_{FG6} + C_{yFG2-yFG5} V_{FG2} + C_{yFG5-yFG8} V_{FG8} + C_{FGCG}\left(V_{CG2} + V_{CG8}\right)}{C_{TUN} + C_{ONO} + C_{xFG4-xFG5} + C_{xFG5-xFG6} + C_{yFG2-yFG5} + C_{yFG5-yFG8} + C_{FGCG}} \tag{5.11}$$

where V_{FG4}, V_{FG6}, V_{FG2}, and V_{FG8} are the floating-gate voltage at cells 4, 6, 2, and 8, respectively, and V_{CG2} and V_{CG8} are the control-gate voltage at cells

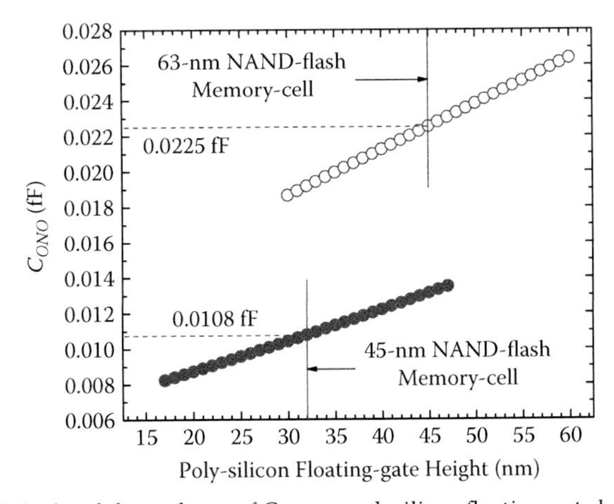

FIGURE 5.23 Calculated dependence of C_{ONO} on polysilicon floating-gate height.

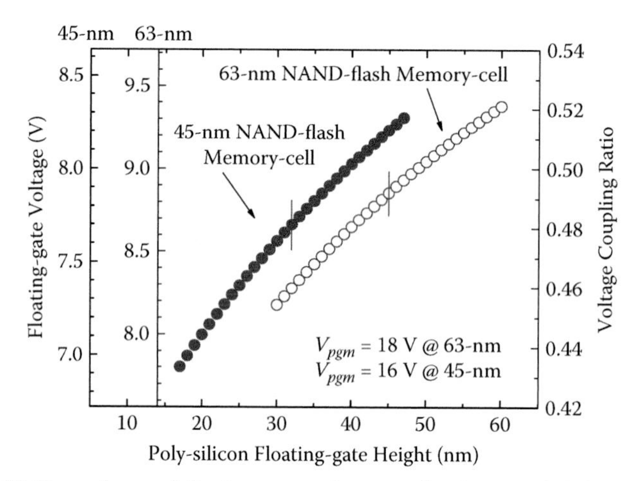

FIGURE 5.24 Dependence of floating-gate voltage on floating-gate height with program voltage biasing.

2 and 8, respectively. For the 63-nm cell (63-nm gate length and 80-nm gate width), the distances between the floating-gates along the word line and bit line are 63 and 50 nm, respectively. For the 45-nm one (45-nm gate length and 57-nm gate width), they are 45 and 50 nm, respectively. Programming voltages were assumed to be 18 and 16 V for the 63- and 45-nm cells, and the pulse duration was 200 μsec. Figure 5.24 shows V_{FG5} as a function of polysilicon floating-gate height determined by the PV of the remaining polysilicon thickness after polysilicon CMP. V_{FG5} increases with polysilicon floating-gate height. The voltage coupling ratio for the 63-nm cell (at 45-nm floating-gate height) is approximately 2.2% larger than that for 45-nm one (at 32-nm floating-gate height). This indicates that the floating-gate voltage coupling becomes weaker as memory size becomes smaller because of the reduction of floating-gate voltage coupling and the enhancement of floating-gate interference.

Fowler–Nordheim tunneling-current density, well known as the programming mechanism of a NAND-flash memory-cell, is represented by

$$J = \frac{q^2E^2}{16\pi\hbar\phi_{ox}}\exp\left[\frac{-4\sqrt{2m^*}\left(q\phi_{ox}\right)^{3/2}}{3\hbar qE}\right] \tag{5.12}$$

where $\hbar$ is the conduction barrier height between SiO_2 and silicon substrate and m^* is the effective mass of electron. The size of the active area generating Fowler-Nordheim tunneling is $W \times L = 51.4 \times 63$ nm for the 63-nm NAND and 36.7×45 nm for the 45-nm NAND. The number of electrons can be approximately obtained because the current density is converted to charge when the pulse duration is 200 μsec, as assumed earlier.

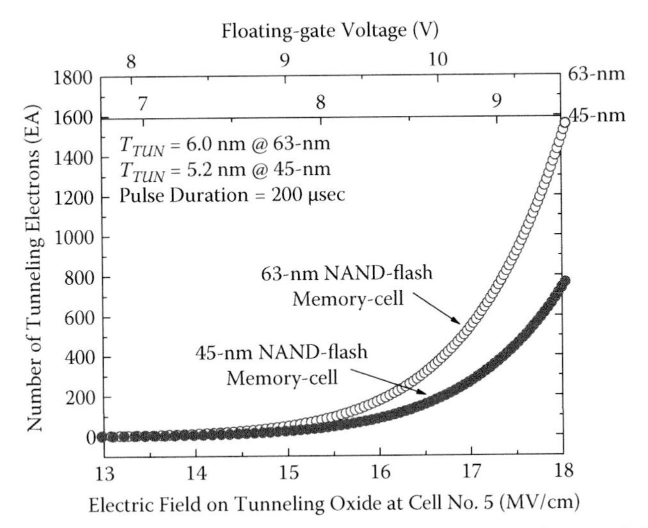

FIGURE 5.25 Number of tunneling electrons versus floating-gate voltage and electric field on tunneling oxide at cell No. 5.

The result is shown in Figure 5.25, where the x-axis is the voltage applied to the tunneling oxide and electric field, and the left y-axis is the number of tunneling electrons. Although, the number of tunneling electrons in the 63-nm NAND is larger than that of 45-nm NAND-flash memory-cell, V_T variation of the latter is severe. This is because the tunneling oxide capacitance of the 63-nm NAND is 1.99 times bigger than that of 45-nm NAND, due to the large active silicon area.

As V_{FG} is changed with floating-gate height, the variation of V_T defined by ΔV_T is represented by

$$\Delta V_{T,63nm} = \frac{\Delta Q_{FN}\left(= Q_{FN,45+\alpha nm} - Q_{FN,45-\alpha nm}\right)}{C_{TUN,63nm}} \tag{5.13}$$

$$\Delta V_{T,45nm} = \frac{\Delta Q_{FN}\left(= Q_{FN,32+\alpha nm} - Q_{FN,32-\alpha nm}\right)}{C_{TUN,45nm}} \tag{5.14}$$

where α is the height of the changed floating-gate. For instance, if the height of the floating-gate has the variation of 10 nm, then α is 5 nm. Therefore, ΔQ_{FN} is $Q_{FN,50nm} - Q_{FN,40nm}$. The variation of V_T classified by thickness, calculated earlier, is represented by the function of the PV of the polysilicon floating-gate height induced by wafer nanotopography, as shown in Figure 5.26. The final V_T variation represents the difference in the variation between two cells that have different polysilicon floating-gate

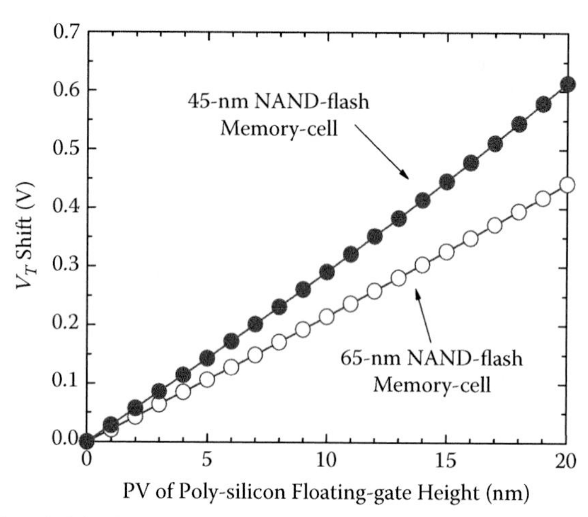

FIGURE 5.26 Threshold voltage of NAND-flash cell shift due to film thickness variation.

heights. Thus, for both the 63- and 45-nm NAND-flash memory cells, V_T variation linearly increased with the PV of the polysilicon floating-gate height induced by wafer nanotopography. However, the V_T variation for the 45-nm cell is higher. The PV of the polysilicon floating-gate height and the PV of the wafer nanotopography are related, as shown in Figure 5.21c. Accordingly, the V_T variation linearly increases with the PV of the wafer nanotopography, as shown in Figure 5.27. The impact of the V_T variation for the 45-nm NAND is higher than that for 63-nm. This shows that the larger the direct influence of the wafer nanotopography, the larger the V_T variation becomes. It is intuitive that the impact of the wafer nanotopography on the V_T variation becomes larger as the memory-cell-device design rule (gate length) becomes smaller. Thus, optimizing silicon-wafer fabrication processes to reduce the PV of wafer nanotopography, and thereby minimize the V_T variation, is key to improving device yield.

5.4 Equipment in Measuring the Nanotopography

Three kinds of instruments for the characterization of nanotopography, namely, SQM, NanoMapper, and DynaSearch are reviewed and compared. The calibration result using identical samples is also shown.

5.4.1 Introduction to General Equipment Used in the Measurement of Nanotopography

How is nanotopography inspected and sorted?

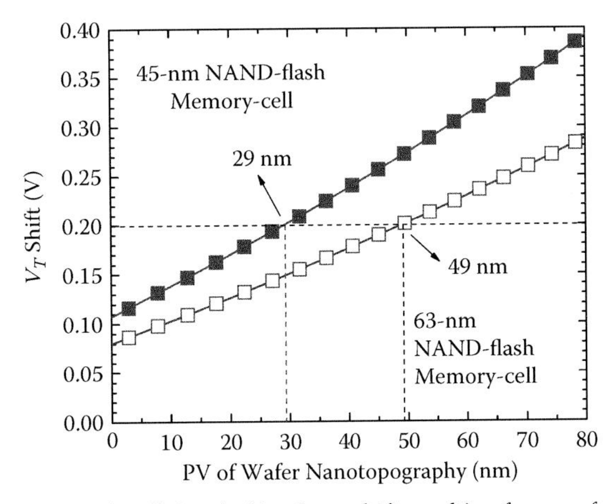

FIGURE 5.27 NAND-flash cell threshold voltage shift resulting from wafer nanotopography.

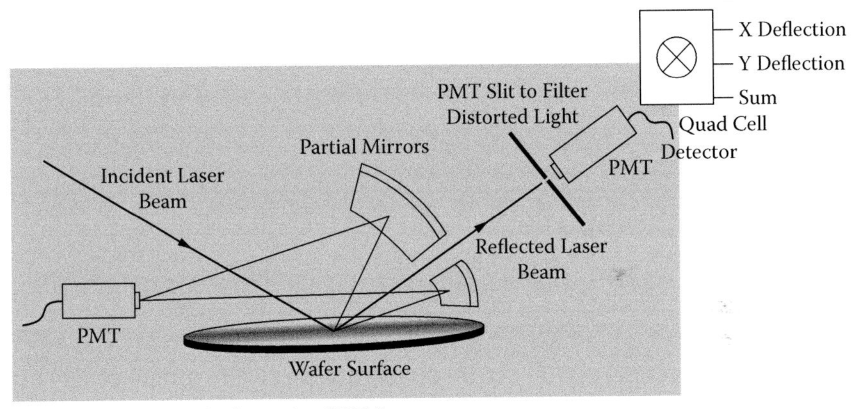

FIGURE 5.28 Optical schematic of SQM.

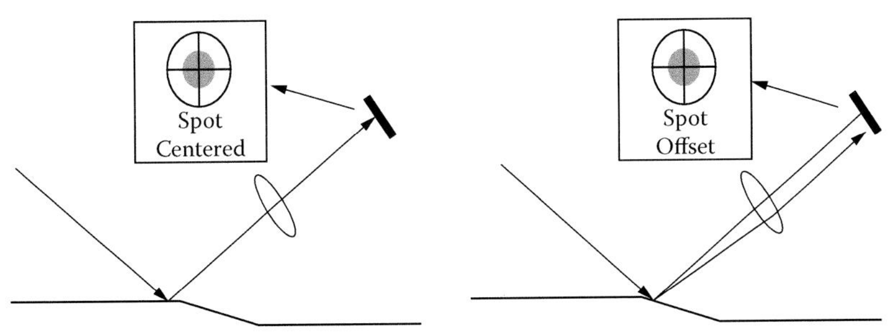

FIGURE 5.29 Quantitative measurement of surface slope.

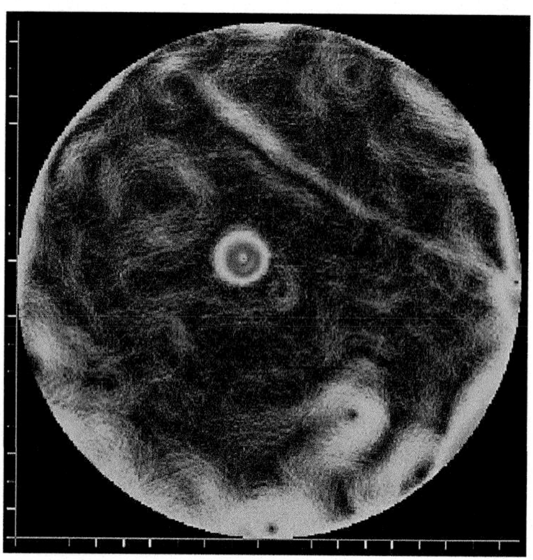

FIGURE 5.30 Slope magnitude map of nanotopography of polished 200 mm wafer.

5.4.1.1 SQM™ (Surface Quality Monitor), from ADE, USA

A. Laser beam reflectance-based bright channel technology (quad cell)

1. AWIS-SQM: It is viewed as a next-generation multifunctional inspection tool; can be used in a 200 mm or 300 mm wafer; has multifunctional inspection capability; and is mainly used in measurement of particle, haze, and nanotopography.
2. CR83-SQM: Currently used as the standard wafer substrate inspection tool worldwide. The WIS CR83-SQM provides wafer manufacturers the required nanotopography measurement capability for design rules down to 0.18 micron and better. It is the production tool for nanotopography inspection, sorting and surface mapping. Utilizing ADE's Quad Cell technology, the WIS CR83 provides height maps based on slope measurements and has a dynamic range from 0.5 mm to 10 mm. Its versatile software allows users to define four unique nanotopography bins for automated sorting by specifying a designated height change (nm) over a specified distance (spatial wavelength). It also allows users to view height maps at five selectable view scales. With these capabilities the WIS CR83 enables wafer manufacturers to simultaneously inspect for nanotopography features and monitor their production processes.

B. SQMTM Technology: Bright Field Quad Cell

- Surface nanometer scale topography height variation and location over whole wafer surface
- Automated sort on user-defined parameters

- Change in reflected angle = 2 × Change in surface slope
- Change in spot location = Focal length × Reflected angel
- Quad cell circuit converts spot location to voltage
- Extract surface slope distribution within a wafer
- Height map is calculated from the integration of the slope

5.4.1.2 NanoMapper, from ADE Phase Shift, USA

A. NanoMapper®: NanoMapper is a nanoscale science and engineering (NSE) knowledge mapping system developed by the Artificial Intelligence Lab at the University of Arizona. NanoMapper enables users to search for patents (1976/1978–2006) or grants (1991–2006) by patent or grant number, keywords, and other data fields. NanoMapper also provides analysis tools. The National Science Foundation supported this research project. NanoMapper is an automated, precision surface mapping system available for research, analytical, and process control applications for 200 mm and 300 mm wafers. Using proprietary, optical interferometry from ADE Phase Shift, the system characterizes polished wafer surfaces by providing whole wafer topology measurements.

   1. Product highlights
      - Wafer nanotopography measurement
      - Measurement features:
        - Able to accurately repeat measurements
        - Measures 200 and/or 300 mm wafers, based on configuration
        - User editable recipes
      - Software features:
        - Measures nanotopography to the edge of the wafer
        - User editable recipes

B. NanoMapper® FA (Figure 5.31): NanoMapper FA has all the analysis options of its R & D brother, but features full factory automation, including FOUP, SMIF, and open cassette capabilities. With NanoMapper FA, you can have: full edge-grip capability, enhanced chuck performance, and additional SECS-GEM functionality. NanoMapper FA is an automated, precision surface mapping system available for research, analytical, and process control applications for 200 mm and 300 mm wafers. Using proprietary, optical interferometry from ADE Phase Shift, the system characterizes polished wafer surfaces by providing whole wafer topology measurements.

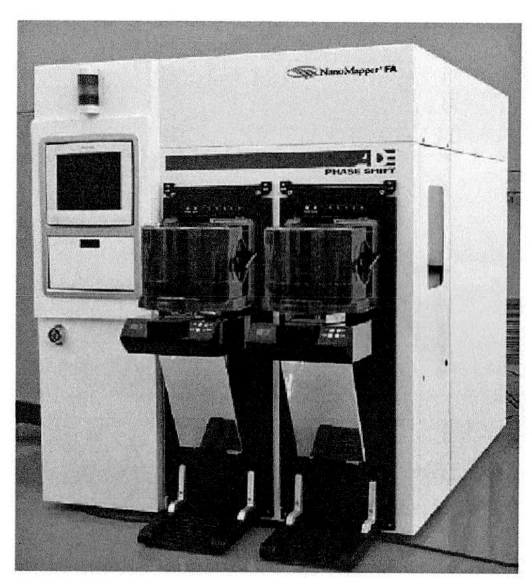

FIGURE 5.31 NanoMapper FA.

1. Product highlights
   - Automated wafer nanotopography measurement
   - Measurement features:
     - Able to accurately repeat measurements
     - Subnanometer height resolution
   - Software features:
     - Measure nanotography to the edge of the wafer
     - Single-statistic pass/fail threshold curve analysis
     - User editable recipes

C. Summary

1. Unfiltered height data typically includes ~10 microns of topography for quality prime wafers.
2. Spatial filtering removes the high amplitude long wavelength shape information to reveal nanotopography.
   - Spatial filter cutoff wavelength should be approximately twice the CMP length.
   - Nanotopography directly impacts post-CMP film thickness variation.

Measuring surface height directly, NanoMapper provides the subnanometer sensitivity necessary to address the process development needs for leading edge semiconductor device design rules down to 0.1 micron. Interactive 3-D graphics and powerful analysis software allow rapid visualization and quantification of polishing process effects. These effects include nanotopology defects that ultimately limit wafer usability during semiconductor device fabrication. The result is faster process development and precision process control during production, with reduced wafer scrap costs.

5.4.1.3 DynaSearch, from Raytex, Japan

DynaSearch (Figure 5.32) is an optical measurement system for evaluating the flatness of 200 mm and 300 mm wafers based on a unique, proprietary image processing algorithm. It views the entire surface of the wafer and gives real-time image data to derive the wafer topography from the angular components of the skewed surface. The system is capable of evaluating both the wafer flatness and nanotopography (Figure 5.33).

Using Raytex's proprietary optical method, the DynaSearch performs scientific wafer topography measurement and inspection to quantify wafer flatness and topography. This one unit handles both the wafer flatness and the nanotopography measurements necessary to evaluate wafer-manufacturing quality. The Raytex's EdgeScan edge inspection, BackScan backside inspection, and flatness and topography measurement performed by the Raytex DynaSearch are a winning product combination that enables users to create a total shipping inspection line.

1. Features:
   - Performs high-resolution high-accuracy wafer topography measurement using Raytex's proprietary optical measurement system

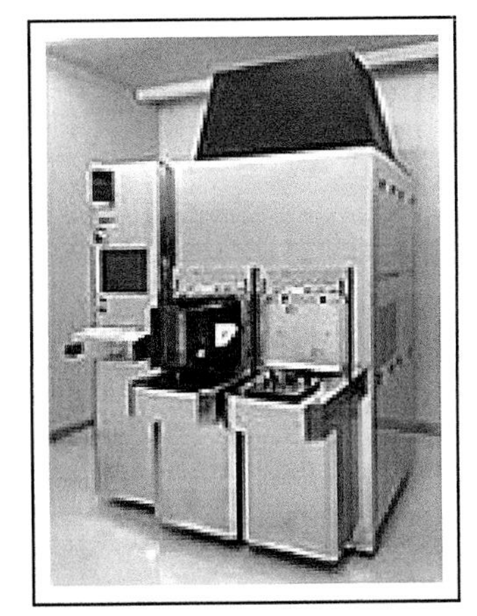

FIGURE 5.32 (a) DynaSearch N4-800, (b) DynaSearch N4-1200.

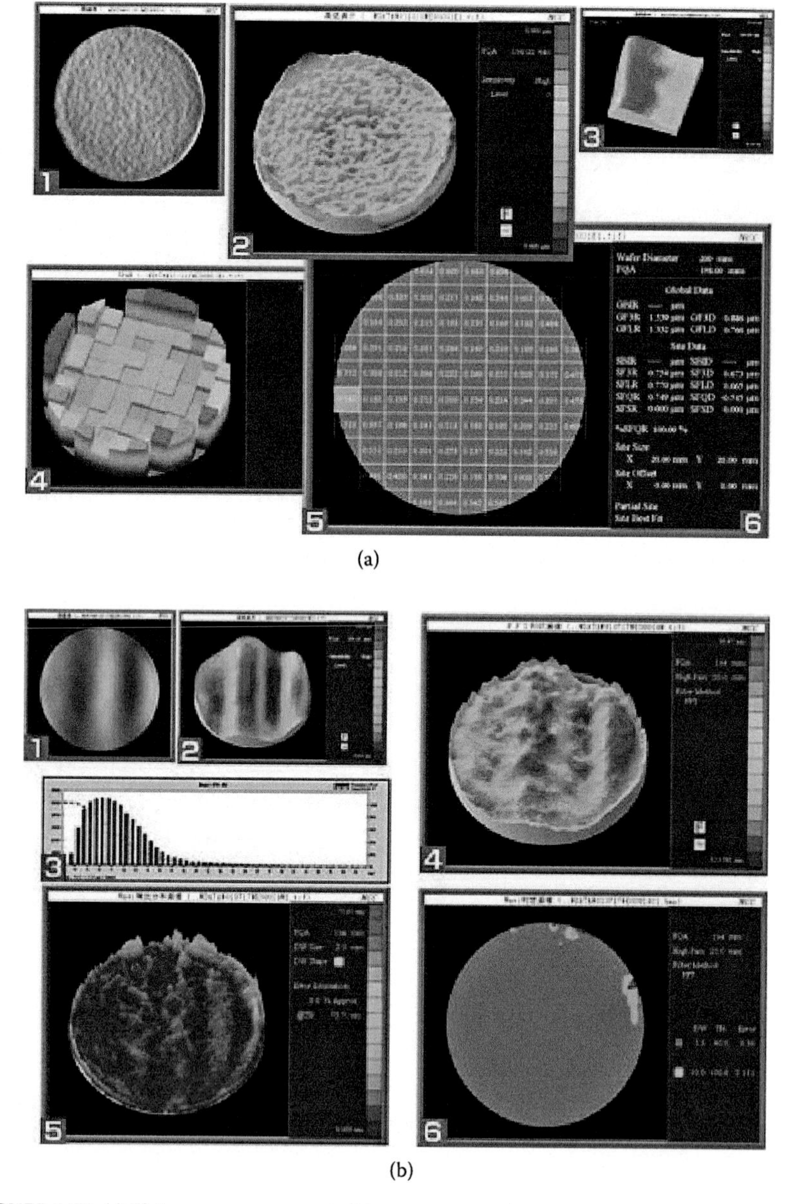

FIGURE 5.33 (a) Flatness measurement, (b) nanotopography measurement.

- Minimal impact due to vibration or other equipment environmental conditions unavoidable using conventional interferometry
- Employs a stepper-type chuck for measurement to reproduce lithography flatness conditions in measurement

2. Measurement items:
    - Flatness measurement
    - Nanotopography measurement
3. Applications:
    - Outgoing/acceptance inspection of mirror and epitaxial wafers
    - Wafer shape evaluation before and after film deposition and epitaxial processes
    - Yield improvement of photolithographic process
    - Conditioning of grinding, polishing, diffusion, and epitaxial growth

5.4.1.4 Line Profile Comparison among Three Instruments

The line profiles of height change along the *x*-axis of the wafer measured by SQM, NanoMapper, and DynaSearch are superimposed on Figures 5.34a–e. Each of five measurements is done for different wafers. The filters applied to raw profiles are "Standard Filter" for SQM and "Gaussian Filter" for NanoMapper and DynaSearch. The cutoff length is set to 20 mm for all methods. From the overview of the profiles, the positions of local peak and valley almost coincide for the three tools. However, SQM gives somewhat different profiles from the other two methods.

5.4.1.5 Calibration among the Standard Deviations of Height Change Measured by Three Kinds of Instruments

Figures 5.35a–c is the calibration among the standard deviations of height change profiles. As for the standard deviation, SQM gave smaller values and variations than NanoMapper or DynaSearch (see also Table 5.1).

FIGURE 5.34 Sensitivity comparisons of SQM, NanoMapper, and DynaSearch: (a) Measurement #1, (b) Measurement #2, (c) Measurement #3, (d) Measurement #4, (e) Measurement #5.

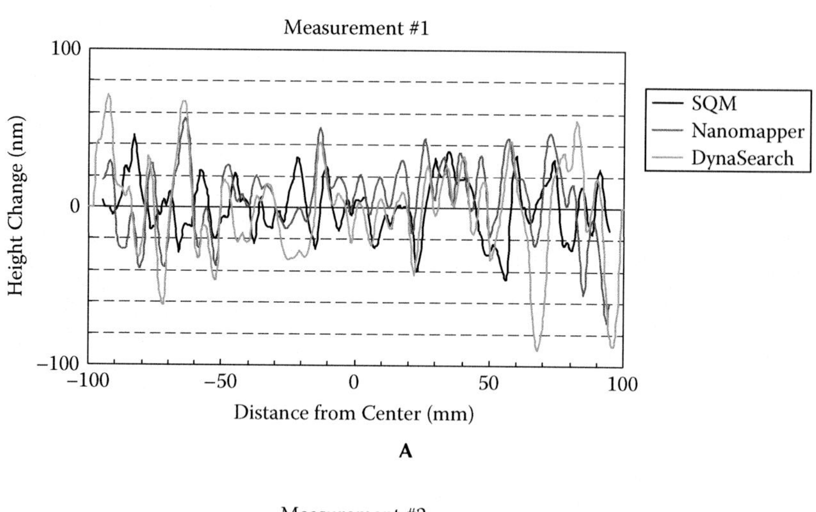

A

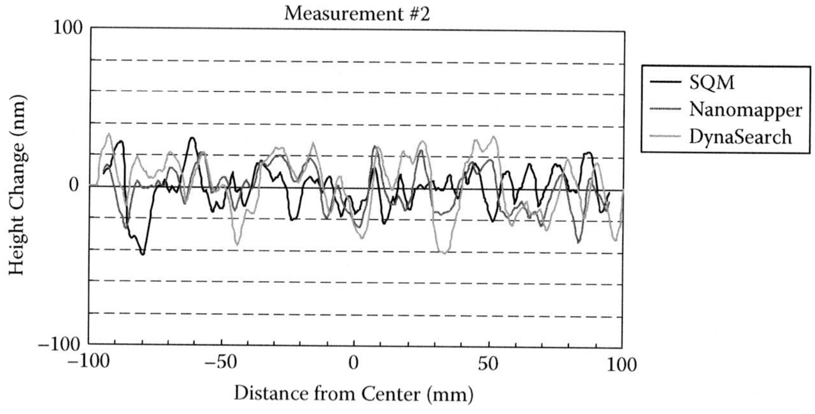

B

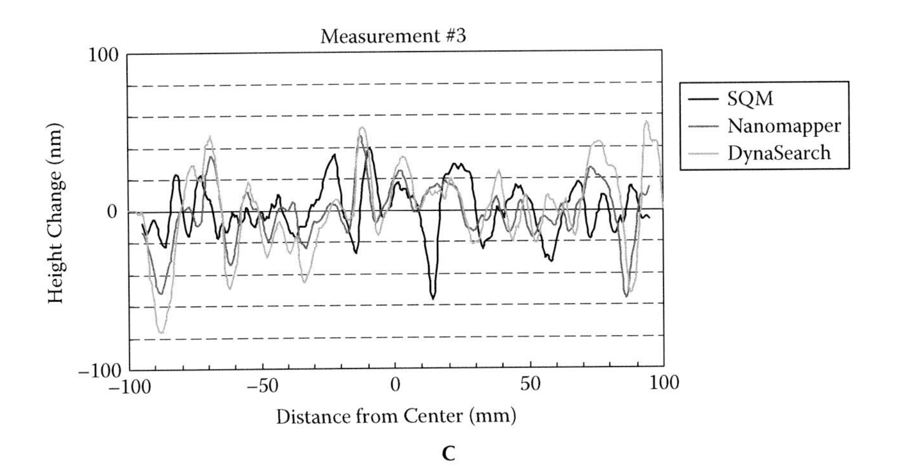
Measurement #3
100
0
−100
Height Change (nm)
−100
−50
0
50
100
Distance from Center (mm)
SQM
Nanomapper
DynaSearch

C

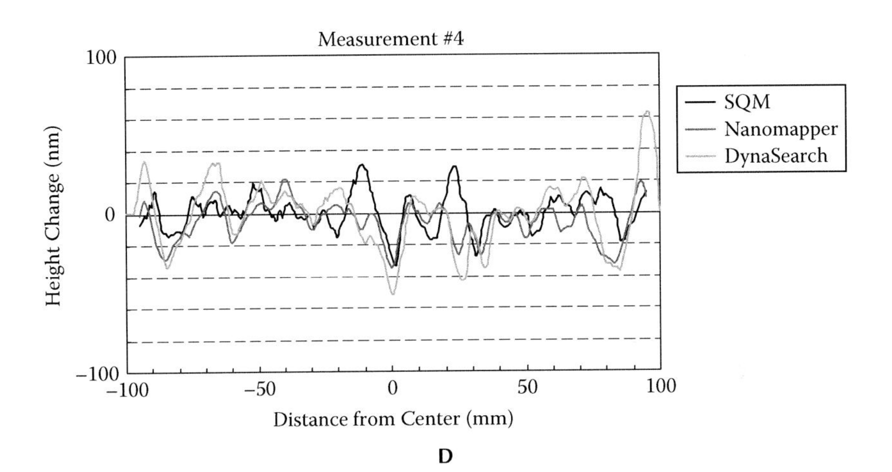
Measurement #4
100
0
−100
Height Change (nm)
−100
−50
0
50
100
Distance from Center (mm)
SQM
Nanomapper
DynaSearch

D

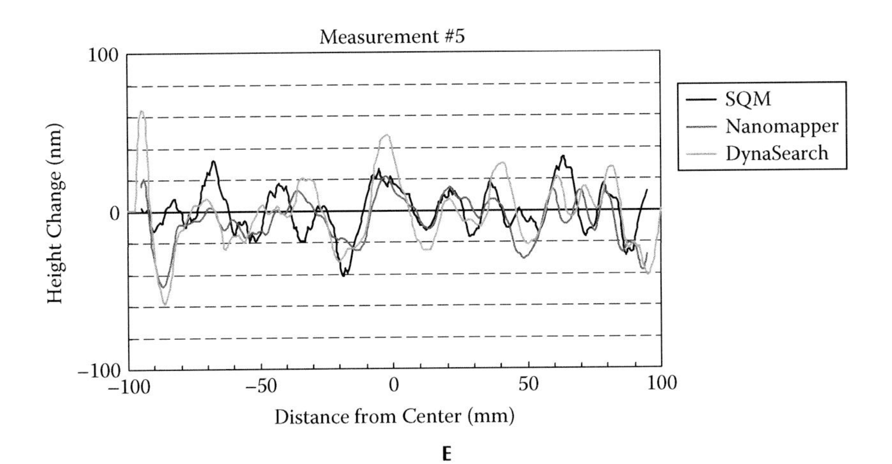
Measurement #5
100
0
−100
Height Change (nm)
−100
−50
0
50
100
Distance from Center (mm)
SQM
Nanomapper
DynaSearch

E

FIGURE 5.35 Standard deviation comparison of SQM, NanoMapper, and DynaSearch: (a) SQM and NanoMapper, (b) NanoMapper and DynaSearch, and (c) DynaSearch and SQM.

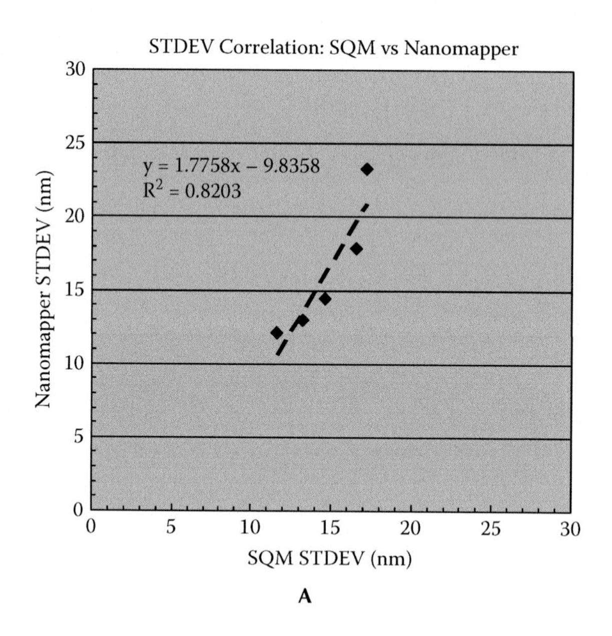

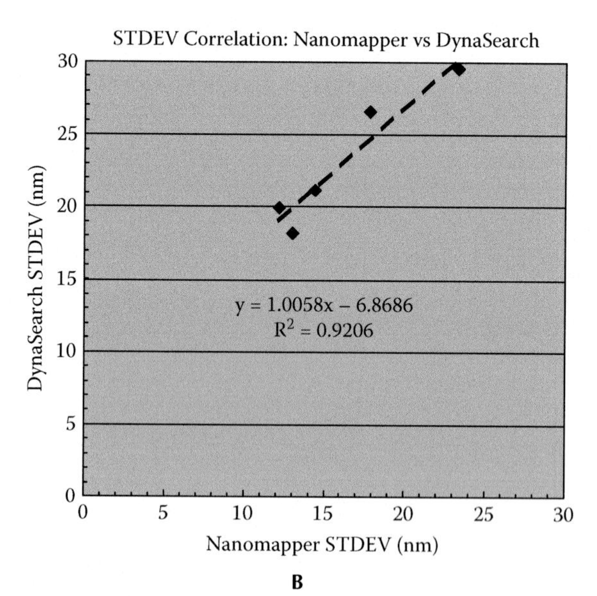

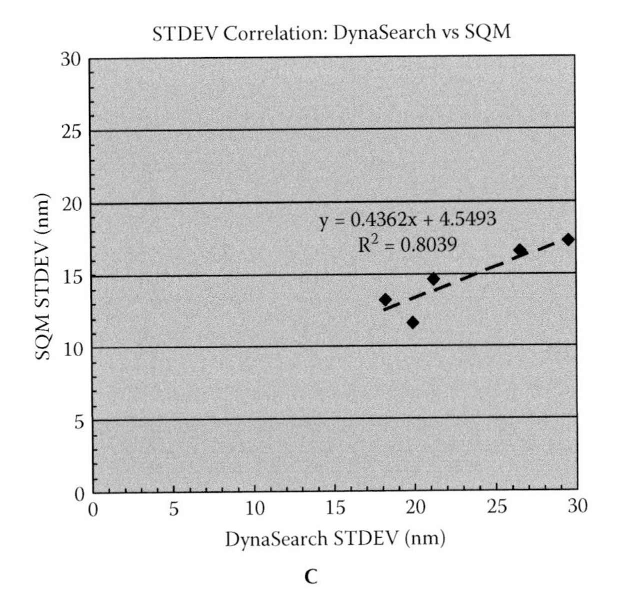
STDEV Correlation: DynaSearch vs SQM
y = 0.4362x + 4.5493
R² = 0.8039
SQM STDEV (nm)
DynaSearch STDEV (nm)
0
5
10
15
20
25
30

C

6

Novel CMP for Next-Generation Devices

6.1 The Progress of Semiconductor Devices upon Current Demand

Chapters 2 through 5 describe CMP process technology for semiconductor device makers. This chapter will describe CMP planarization technology for emerging devices and materials. From the point of view of the researcher whose specialty is detailed (or accurate) processing, the advent of CMP technology has diverse meaning. Silicon wafers on devices are an unapproachable part to the mechanic processing technician. Because of complex factors, the structure of a device's circuit is becoming highly integrated and super miniature. As a result, the degree of rugged processing surface falls under nano scale because device processing and machine processing have very different materials of processing. Devices are shrinking to under 60 nm; therefore, new material and structure processing are appearing. To adapt to this condition, more progressive CMP processing is required and new structures and materials will arrive successively only through CMP processing.

The memory industry is led by metal-oxide semiconductor (MOS) memory, which is largely divided into volatile and non-volatile memory (Figure 6.1). The example of volatile memory is dynamic random access memory (DRAM), whereas non-volatile memory is representative of flash memory. DRAM and flash memory developed rapidly due to the demands for high speed and high capacity devices such as computers, digital cameras, mobile phones, and MP3s (Figure 6.2). New structures and materials were used for improving the performance of these devices. CMP processing has faced a new challenge as well.

Even though the integration of DRAM and flash memory is increasing during this brilliant growth, it is expected that it will hits its limits. To overcome its limit, instead of MOS memory, the latest non-volatile memory research includes phase-change random access memory (PRAM), nano-floating gate memory (NFGM), polymer random access memory (PoRAM), and resistance random access memory (ReRAM). But similar to MOS memory, these new non-volatile memory systems cannot increase integration and form multilayer without planarization processing. This chapter introduces DRAM under 60 nm of MOS memory, CMP processing applied to NAND flash, and CMP processing for next-generation memory producing.

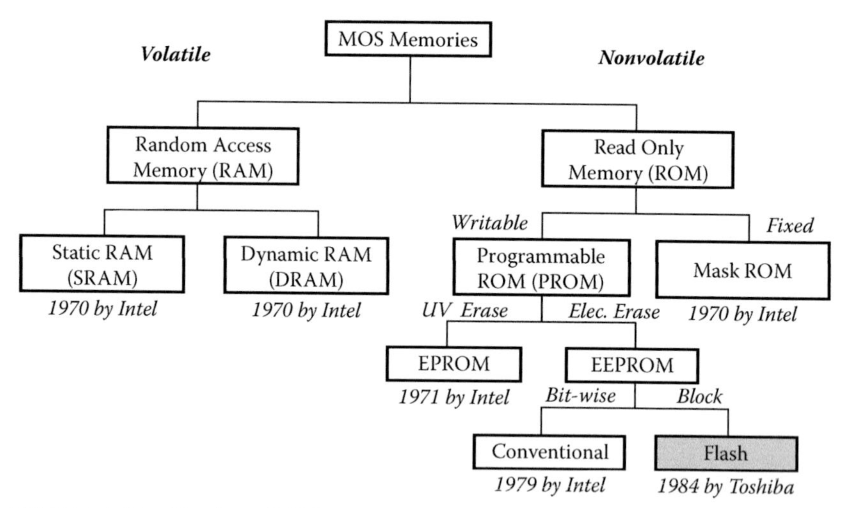

FIGURE 6.1 Classification of MOS memories.

	Year of Production	2005	2006	2007	2008	2009	2010	2011	2012	2013
	DRAM ½ Pitch (nm) (contacted)	80	70	65	57	50	45	40	36	32
	MPU/ASIC Metal 1 (Ml) ½ Pitch (nm)	90	78	68	59	52	45	40	36	32
	MPU Physical Gate Length (nm)	32	28	25	23	20	18	16	14	13
	DRAM Product Table									
	Cell area factor [a]	8	8	8	6	6	6	6	6	6
IS	*Cell area [Ca = af2] (μm^2)*	0.051	0.041	0.032	0.019	0.015	0.012	0.0096	0.0077	0.0061
	Cell array area at production (% of chip size) §	63.00%	63.00%	63.00%	56.08%	56.08%	56.08%	56.08%	56.08%	56.08%
	Generation at produciton §	1G	2G	2G	2G	4G	4G	4G	8G	8G
	Functions per chip (Gbits)	1.07	2.15	2.15	2.15	4.29	4.29	4.29	8.59	8.59
	Chip size at production (mm^2) §	88	139	110	74	117	93	74	117	93
	Gbits/cm^2 at production §	1.22	1.54	1.94	2.91	3.66	4.62	5.82	7.33	9.23
	Flash Product Table									
	Flash ½ Pitch (nm) (un-contacted Poly)(f)	75.7	63.6	56.7	50.5	45.0	40.1	35.7	31.8	28.3
	Cell area factor [a]	4.0	4.0	4.0	4.0	4.0	4.0	4.0	4.0	4.0
IS	*Cell area [Ca = af^2] (μm^2)*	0.023	0.016	0.013	0.010	0.008	0.006	0.005	0.004	0.003
	Cell array area at production (% of chip size) §	67.5%	67.5%	67.5%	67.5%	67.5%	67.5%	67.5%	67.5%	67.5%
	Generation at production § SLC	4G	4G	4G	8G	8G	8G	16G	16G	16G
	Generation at production § MLC	8G	8G	8G	16G	16G	16G	32G	32G	32G
	Functions per chip (Gbits) SLC	4.29	4.29	4.29	8.59	8.59	8.59	17.18	17.18	17.18
	Functions per chip (Gbits) MLC	8.59	8.59	8.59	17.18	17.18	17.18	34.36	34.36	34.36
	Chip size at production (mm^2)§ SLC	144	101.8	80.8	128.3	101.8	80.8	128.3	101.8	80.8
	Chip size at production (mm^2)§ MLC	144	101.8	80.8	128.3	101.8	80.8	128.3	101.8	80.8
IS	*Bits/cm^2 at production § SLC*	3.0E+09	4.2E+09	5.3E+09	6.7E+09	8.4E+09	1.1E+10	1.3E+10	1.7E+10	2.1E+10
IS	*Bits/cm^2 at production § MLC*	6.0E+09	8.4E+09	1.1E+10	1.3E+10	1.7E+10	2.1E+10	2.7E+10	3.4E+10	4.3E+10

FIGURE 6.2 Technical roadmap for DRAM and NAND flash memory.

6.2 Complementary Metal-Oxide Semiconductor (CMOS) Memory

Memory semiconductors should be of high capacity through high-integrated circuits. DRAM cell factor evolved from $8F^2$ into $6F^2$. Along with expectations of 45 nm in 2010 to 32 nm in 2013, there is an anticipation that capacity per chip will increase fourfold like 2.15 Gbit to 8.59 Gbit. Flash memory keeps the design rule of $4F^2$ and when future technology passes through 56.7 nm to 28.3 nm, the capacity per chip will increase rapidly from 8 G to 32 G.

Fulfilling multilayer and miniature structures of memory devices led to the introduction of new materials and structures. For the structure, the design rule decreases less than 70 nm and the short channel effect (SCE) phenomenon appears to have a bad influence on the device drive if existing planar transistor (TR) is applied. To solve this problem, studies are in progress to apply recessed channel array TR and three-dimensional structured FinFET in DRAM and floating gate, twin SONOS, and FinFET SONOS in flash memory (Figure 6.3).

New materials are applied to maximize the capacity of device. To increase the capacitance of cap used in DRAM, studies about high-k dielectric material are in process. Flash memory uses a gate material with polysilicon by reason of high speed and stable storage. To reduce semiconductor device RC delay, Cu metal lines and low-k are being introduced. This section represents the concept of CMP processing being introduced to DRAM and NAND flash devices.

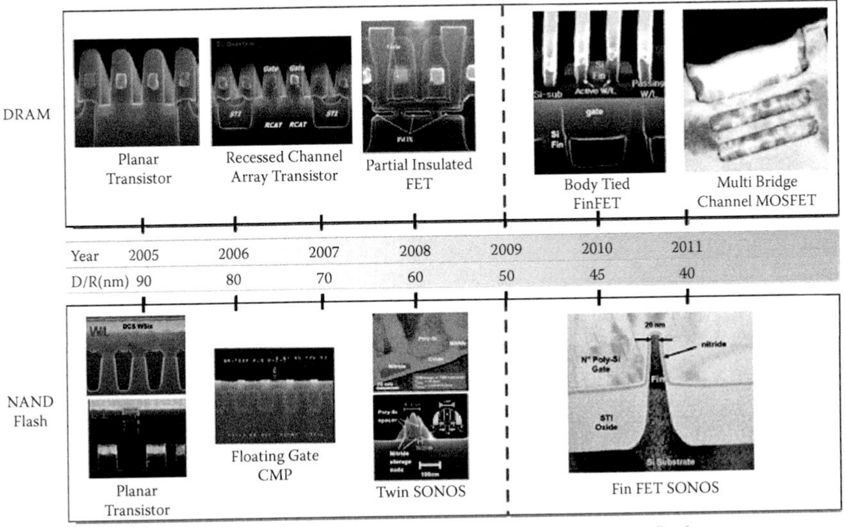

FIGURE 6.3 Roadmap for transistor structures of DRAM and NANA flash memory.

A RCAT structure of three-dimensional recess channel array transistor was developed for the device memory to solve the problem of short channel effect, a demand of integration. The surface of device procession reduces and channel makes recess to lengthen the valid length. As a result, short channel effect can be reduced. Future technology under 60 nm will anticipate the partly insulation FET (PiFET) structure to the part of channel. PiFET that are drawn to the concept of body-tied SOI limits the junction depth by dielectric layer to easily form ultra shallow junction. It also has a small amount of leakage current and it has a merit of SOI that short channel effect is almost nonexistent. This is a device to diminish the production cost. DRAM of future technology under 45 nm is anticipated to have a MBCFET device structure of metal-oxide semiconductor field-effect transistor (MOSFET), which has multichannels by passing through FinFET. This device makes driving current larger and has a great advantage of current resistance capability of a gate because it has a gate all-around (GAA) structure. The self-aligned STI processing, which is a method of formation of floating gate into a device less than 60 nm, enlarges the surface of the floating gate. Therefore, it can increase coupling ratio. To avoid the coupling phenomenon, floating gates to the active area are aligned correctly; that is self-aligned poly Si floating gates fabricated by the CMP process are necessary to overcome the misalignment between the active area and the floating gate in cell arrays.

6.2.1 Noble Metal CMP for DRAM

Existing SiO_2 used in gate dielectric faces its limit because thickness of thin film becomes thinner by the integration of the semiconductor device. Indeed, power dissipation exceeds the standard rate because of the leakage current by tunneling of carrier by electric field is increased. Consequently, it has the same EOT electronically. High-k dielectric makes possible the embodiment of thickness of thin film without tunneling, is physically interested.

With using high-k dielectric film, simplifying the cell structure and formation process is the most efficient method to ensure sufficient capacitance for the high-integrated capacitor's role in narrow surfaces like miniaturizing the next generation of DRAM. Previously used dielectric film in low-k materials are NO of Si_3N_4/SiO_2 and ONO of $SiO_2/Si_3N_4/SiOX$. For the next-generation capacitor, the high-k materials mainly used are Al_2O_3, HfO_2, ZrO_2, Ta_2O_5, BST, and STO, as well as other materials such as HfSiOX and ZrSiOX. In semiconductor industry, high-k thin films form film by chemical vapor deposition (CVD). In relation to this, the study and developments of CVD precursors are as follows.

Capacitor using HfO_2 base dielectric film, applies TiN by top/bottom electrode. However, capacitor technology under 50 nm needs to develop new dielectric material and electrode material. So, noble metals, such as

Ru, Pt, and Ir, are being researched as new electrode materials. In the case of the noble metal, which is a stable material, it is not easy to form a capacitor by using etch back or CMP processing. Noble metal CMP processing, including Ru, should oxidize the surface of polishing target film for the polishing process like other metal CMP.

In Ru CMP, polishing stops when oxide, dielectric between electrodes, exposures while Ru is polished. The removal rate of Ru is under the control of an oxidizer added into slurry. The Pourbaix diagram of Ru shows that RuO_2 and Ru_2O_5 exist when Ru oxidizes, and the removal rate of noble metal CMP depends on the degree of oxidization (Figure 6.4). However, in the real situation, there is no oxidizer that can secure the safety of slurry and strongly oxidize noble metal at the same time. Moreover, for successful Ru CMP, polishing selectivity of Ru:oxide should be considered with STI CMP processing.

In this difficult situation, the reason for applying noble metal CMP is closely related to the electronic characteristics of a device. For example, Ru CMP makes it possible to produce a capacitor to have a higher capacitance than with dry etching because it loses less than dry etching and it has a clear pattern formation as in Figure 6.5. That is, the electronic characteristic can be improved when Ru CMP applies to a device process. In fact,

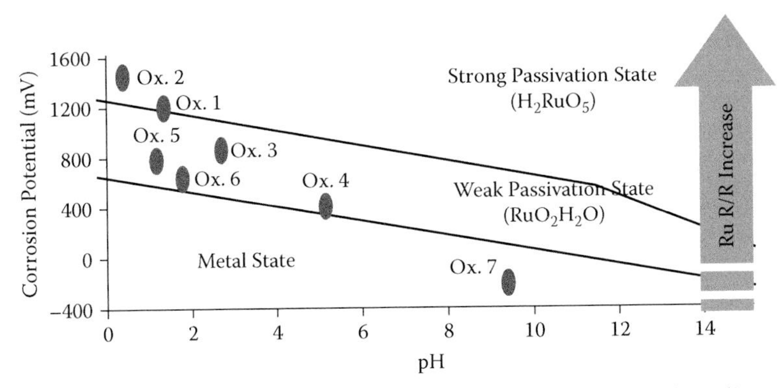

FIGURE 6.4 Electrochemical characterization of various oxidizers in Pourbaix diagram of Ru, obtained from Tafel plots and pH of the slurries.

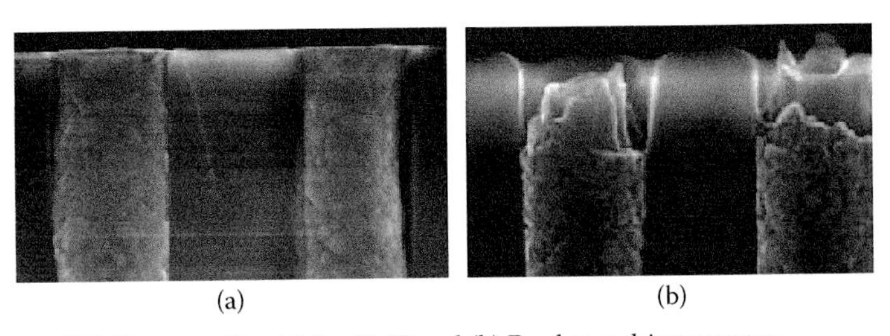

FIGURE 6.5 SEM images after (a) Ru CMP and (b) Ru dry etching process.

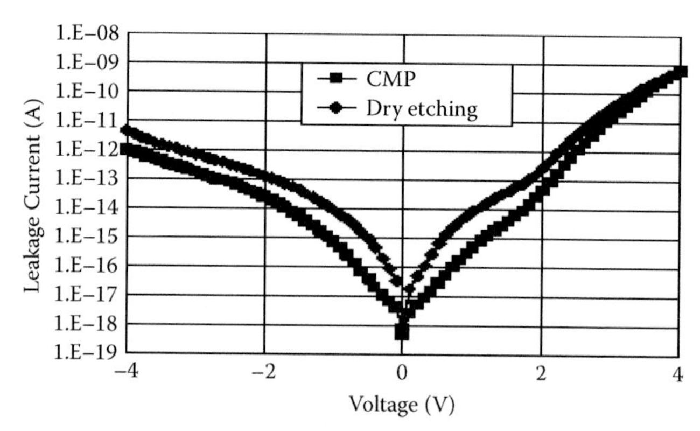

FIGURE 6.6 Comparison of leakage current of RIR capacitors node separated by Ru CMP and Ru dry etching.

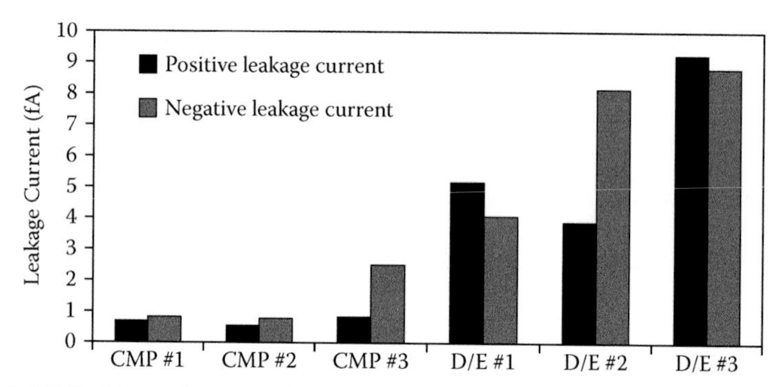

FIGURE 6.7 Positive and negative leakage current of RIR capacitor at 1V and –1V.

when processing by dry etching, the leakage current is 6.537 fA at 1V. As shown in Figure 6.6 and Figure 6.7, the value is remarkably low when a device is processed by CMP. In addition, cell capacitance is 13.4 fF/cell; the CMP process makes it possible to get a much higher value than dry etching, which is 8.4 fF/cell.

6.2.2 Poly Si CMP for NAND Flash Memory

8 G NAND flash memory connects 32 cells in a sequence without source and drain contact each cell. Two transistors of SSL (source select line) and GSL (ground select line) are connected in a series between bit-line contact and CSL (common source line). Floating gate and control gate exist in each cell and is formed channel by controlling the voltage of control gate (Figure 6.8a). Electron that transfers this channel becomes tunneling to be accumulated at the control gate to be a role of memory. The state of stored electron at the floating gate through F-N tunneling is the program state.

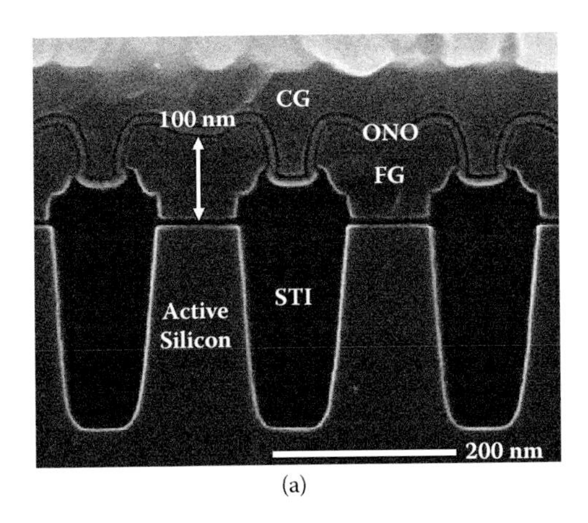

(a)

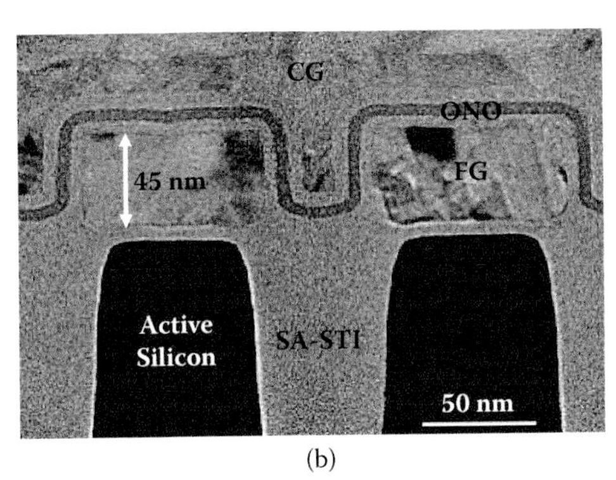

(b)

FIGURE 6.8 (a) Cross-sectional SEM image of a 90-nm NAND-flash memory-cell without CMP. (b) Cross-sectional TEM image of a sub-60 nm structure with interpoly ONO dielectrics.

Missing electron state is the erase. Programming and erasing are formed according to the size of V_{th} of cell transistor.

So, the V_{th} of a device rises at programming time. The state of program/erase can be confirmed through the difference of V_{th}. The augmentation of integration is easily achieved because the simplicity of the structure of NAND flash memory and scaling down are easy. Recently, the emergence of multilevel cell (MLC) makes to have a higher integration. This reduction of design rule might make it possible to be at least 40 nm. Because of the diminution of charge loss tolerance based on scaling down by word-line applied voltage with the decreasing of coupling ratio, interference of floating gate becomes a severe problem. Coupling ratio suddenly decreases at the 60 nm level. Formation processing of the existing floating gate and self-aligned CMP process, like Figure 6.8(b), are brought in because there is no space to increase the size of floating gate that can enlarge the coupling ratio.

With a method of forming floating gate into a 60-nm level device, when polysilicon CMP is used in connection with self-aligned STI process, the surface of floating gate can be extended like in Figure 6.8b. This leads to an increased coupling ratio. Polysilicon CMP is an absolutely necessary process for improving the capacity of flash memory. When slurry, which has high removal selectivity to polysilicon and oxide, makes progress CMP processing, polysilicon recess is occurred by dishing, which is CMP's characteristics. Or electrical characteristics inside cells, including V_{th}, can become worse by thickness distribution of floating gate, which occurs by CMP non-uniformity. The voltage of floating gate is capacitance based on ONO (oxide–nitride–oxide) between control and floating gate. $C_{Tot}(= C_{ONO} + C_{Tunnel})$ is capacitance based on Tox and capacitance based on ONO is decided according to the voltage of control gate.

To obtain a high coupling ratio, the value of ONO capacitance should be increased. Capacitance needs to maximize the selective surface above floating gate. But the width of floating gate, which is a factor for deciding a selective surface, is determined when the design rule was set. The height of the floating gate will be a key factor to determine the coupling ratio in the end.

Furthermore, stored electron at the floating gate rapidly decreases in accordance with the decreasing of design rules. In case of SLC (single-level cell), electron loss less than 20% is allowed within the same range. In case of MLC, each program state has to have the same V_{th} range. To reduce the tolerance limit of electron loss per each state, control gate and capacitance of floating gate should be increased; however, scaling down of interpoly ONO ends its limit. Another problem is that the gap between cells becomes narrow according to the increasing of cell integration, and an interference phenomenon will arrive by capacitive coupling between floating gates. The transition of cell V_{th} by interference of floating gate brings up the V_{th} transition of around cell. So, when V_{th} changed in the phase of programmed one cell, V_{th} is changed due to the coupling phenomenon in which electrode is accumulated in floating gate by writing to the next cell. This change of V_{th} becomes a severe problem at the MLC action, which should keep the V_{th} gap between cells. Voltage of floating gate is influenced by control gate of around cell and floating gate, and is not only influenced by the voltage of control gate of the corresponding cell.

As described earlier, by increasing integration, coupling ratio of floating gate, permitted possible charge loss, and V_{th} shift by capacitive coupling of intercells become problems. Especially, due to narrowing the space between bit line and word line, V_{th} shift of cell dramatically increases around 0.2 V to make higher V_{th} distribution through interference of floating gate between cells. To decrease this interference, the space between cells fills with low-k material or reduces the height of floating gate. At the same time, exact thickness control of floating gate should be needed.

Especially, the height of floating gate should be controlled through the CMP process because that greatly influences word-line and bit-line cells.

To achieve the isolated poly floating gate, the polishing should be stopped at the oxide film. Therefore, the poly-to-oxide removal selectivity is the most important factor for the poly isolation CMP process. Poly isolation CMP without selectivity induces dishing and rugged topography on surfaces, which result in deterioration in the quality of interlayers in the device.

The two steps of poly Si CMP are shown in Figure 6.9. The first step is to remove Si, then polishing at oxidize film is to be stopped in the second step. The polishing of Si needs an addition for accelaration polishing because it should be progressed over 2000 Å per minute. This additive is mainly a chemical compound that has an amine group and abrasives polish the Si film by the complex formation of Si surface and amine group. For the last step, lower the polishing rate of oxidized film to less than 100 Å/min for stopping polishing.

The difference of hydrophobicity between poly Si and oxide film is the key factor to achieving the high removal selectivity in the poly isolation CMP process. As oxide film is more hydrophilic than silicon film, hydrophilic polymer is preferred to be adsorbed on the hydrophilic oxide film. By utilizing selective adsorption of hydrophilic polymer, the passivation layer formed on oxide film can prevent the direct contact of abrasive particles, which results in suppression of the removal rate of oxide film during poly isolation CMP process (Figure 6.10).

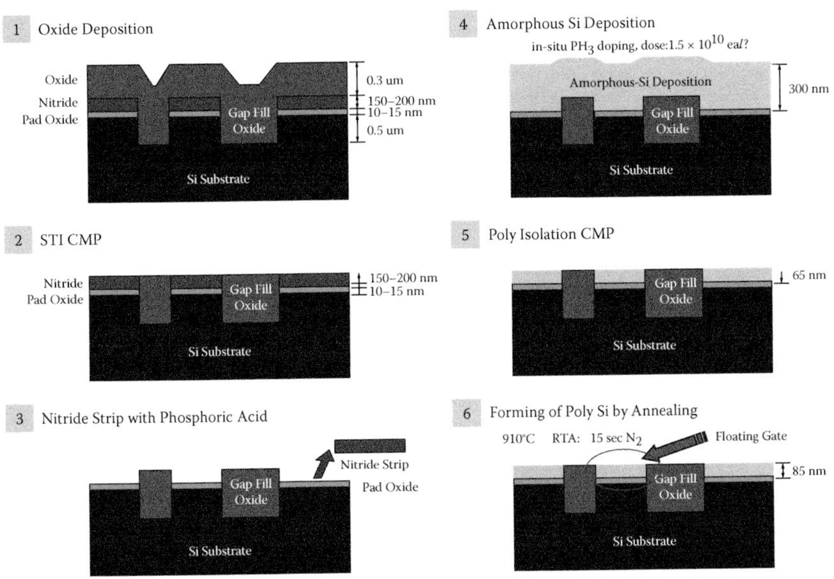

FIGURE 6.9 (See color insert) Schematic process flow of the poly isolation CMP process.

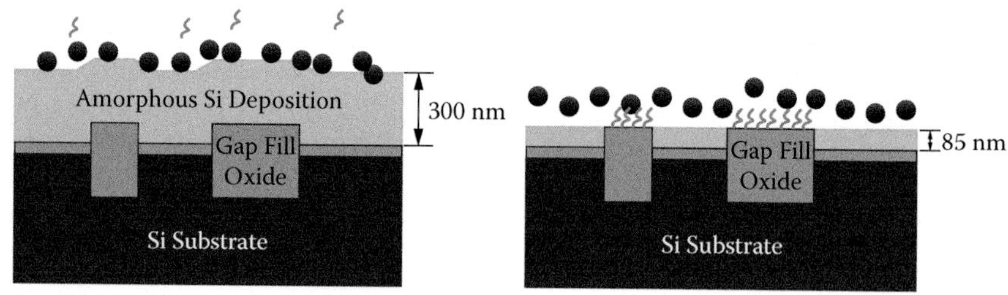

FIGURE 6.10 (See color insert) Mechanism for the poly isolation CMP process.

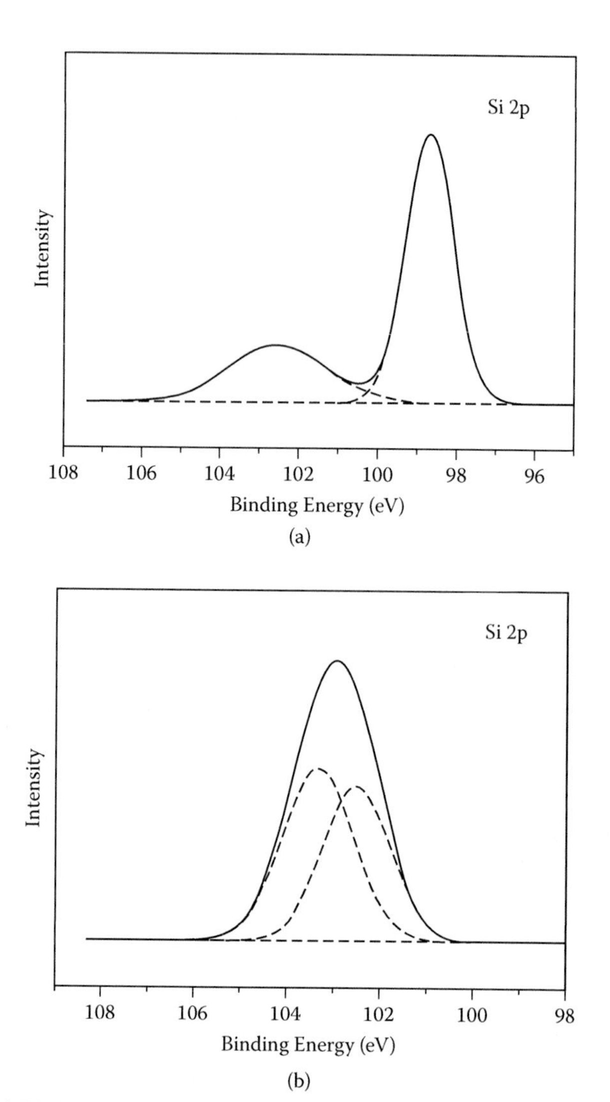

FIGURE 6.11 XPS spectrum of poly Si and oxide film at pH 10.

To investigate the surface structures of poly Si and the plasma-enhanced tetraethylorthosilicate (PETEOS) films, Si 2p XPS analysis for each film was conducted as shown in Figure 6.11. The Si2p spectra of poly Si film in Figure 6.11a shows two symmetrical profiles, of which binding energies correspond to 98.7 eV for Si and 102.5 eV for SiOH. Compared with the intensity of Si peak, that of SiOH is considerably low, which mean the surface structure of poly Si film consists of Si structure for the most part. On the other hand, as shown in Figure 6.11b, the spectrum observed from the PETEOS film was mainly separated into two symmetrical profiles. The binding energies observed from this result are 102.50 and 103.30 eV, which correspond to SiOH and SiO, respectively. The intensities of the two peaks are almost similar and this observation indicates that a respectable amount of siloxane group is present on the surface of PETEOS film.

The difference originated from hydrophilic and hydrophobic characteristics is confirmed from contact angles of deionized (DI) water on each surface. The contact angle of DI water on poly Si film was 61.2° in the Kruss contact angle measurement system. On the other hand, DI water drop exhibited a low contact angle below 10°, thus spread out on the oxide wafer as shown in Figure 6.12. This result is coincided with the XPS result; that is, the surface of SiO_2 is more hydrophilic than that of poly Si due to siloxane (≡Si–O–Si≡) bonding.

Figure 6.13 shows the adsorption isotherms for PAM on poly Si and SiO_2 as a function of PAM concentration. Adsorption of PAM on oxide surfaces increases and reaches a plateau level of approximately 0.23 mg/m^2. However, PAM is scarcely adsorbed on poly Si surfaces. This is driven by the difference in hydrophobicity, which affects the interaction between PAM and each surface. At high pH, the negative site MO^- of metal oxide (MO) surface bonds with the weakly acidic NH_2 function. Therefore, the interaction between SiO^- of SiO_2 surface and NH_2 group of PAM led to the selective adsorption of PAM on SiO_2.

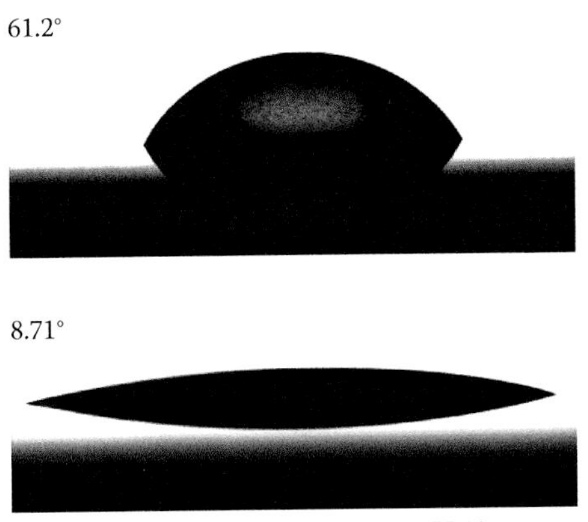

FIGURE 6.12 Contact angle of poly Si and oxide film at pH 10.

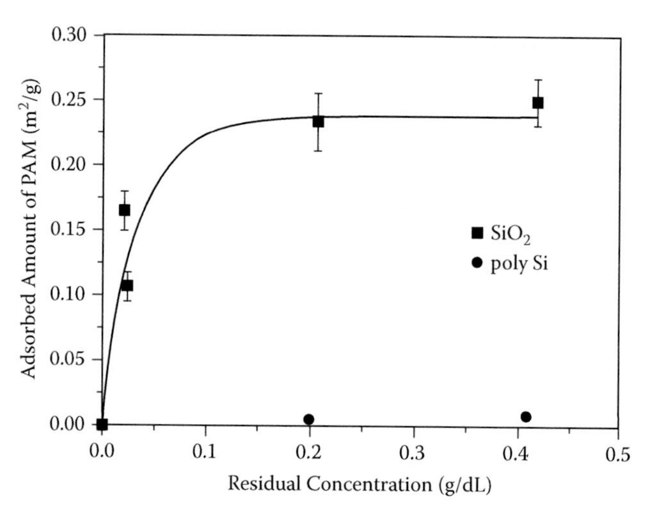

FIGURE 6.13 Adsorption isotherms of PAM on poly Si and oxide at pH 10.

Adhesive forces were measured to analyze the interaction between Si atomic force microscope (AFM) tips and the deposited films from the force-distance measurement with the AFM. The force-distance curve is expected to elucidate the adsorption behavior of PAM on the deposited films. In the force-distance measurement, there is no interaction until the tip is close enough to be attracted to the surface. As the tip approaches to the surface, it contacts with the surface. After the contact, the tip is retracted from the surface, the cantilever is bent, and a repulsive force (positive) is measured. When the tip is being retracted, an attractive force is measured (negative). When the critical force is reached, the tip is separate from the surface and this point is called the pull-off point. Therefore, the pull-off point, which corresponds to the point of the critical force, is determined by the degree of the adhesive force between the tip and the surface. The higher the adhesive force between the tips and the films, the lower the pull-off point. Figure 6.14a shows the force–distance curves of the tips and poly Si film at pH 10 as a function of the PAM concentration. In the absence of the absorbed PAM molecule, an adhesive force was observed at approximately 20 nm of separation distance. It is of interest that there is no significant difference between the surface forces of the tip and poly Si film even with the presence of PAM. This result is almost the same for all samples, irrespective of the concentration, which means that PAM is scarcely adsorbed on poly Si film.

On the other hand, Figure 6.14b shows the force–distance curves of the tips and the oxide film at pH 10 as a function of the PAM concentration. In the absence of the absorbed PAM molecule, an adhesive force was observed at approximately 30 nm of separation distance. However, it was found that the adhesive force disappeared by the addition of PAM due

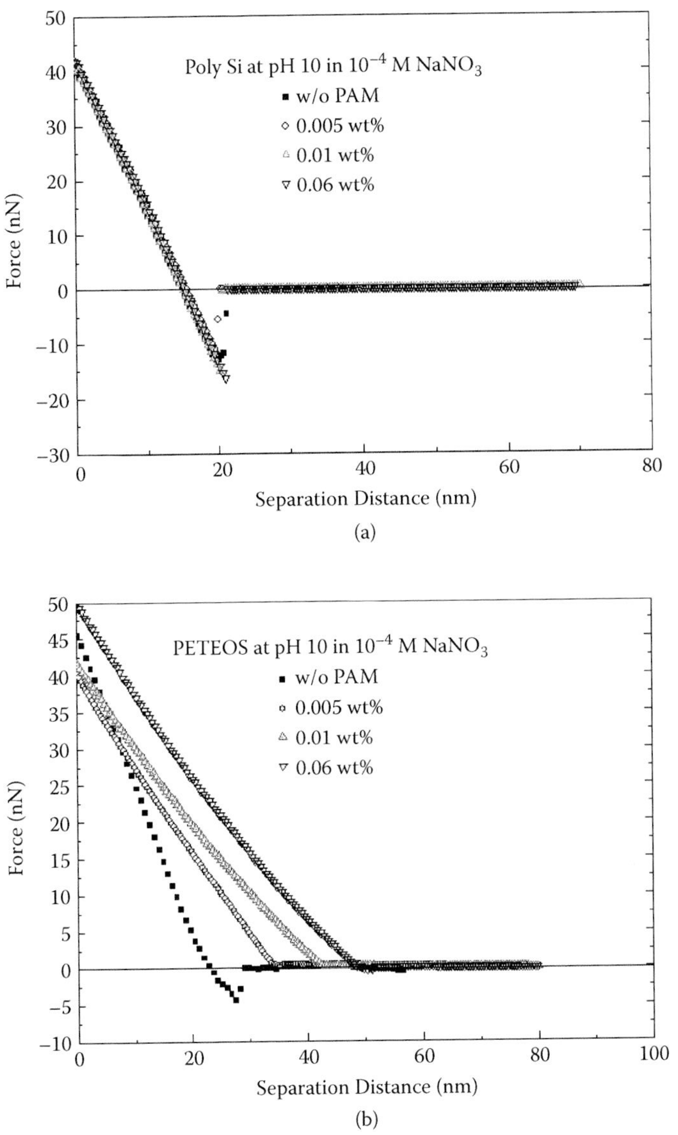

FIGURE 6.14 Force-distance curve between the AFM tip and the surface of wafer: (a) poly Si, (b) oxide.

to the interaction between the siloxane bonding of the oxide film and the amine group of PAM at pH 10. This result is more clearly observed as the concentration of PAM increases. Therefore, it is expected that the presence of PAM causes the suppressed removal rate of the PETEOS film by the formation of the passivation layer on the oxide film and the increase in the concentration of PAM leads to the decrease in the oxide removal rate.

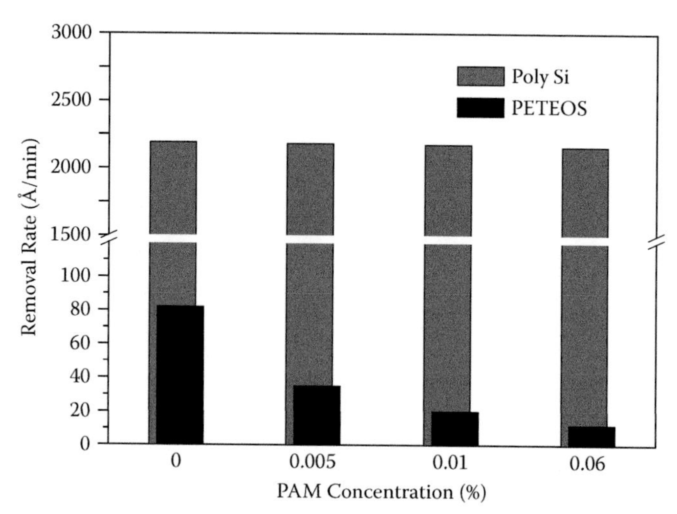

FIGURE 6.15 Removal rate of poly Si and oxide film.

The results of the CMP performance as a function of the concentration of PAM are shown in Figure 6.15. As the concentration of PAM in poly isolation CMP slurry increased, the slurry for poly isolation CMP process had a higher removal rate for the PETEOS film, reduced from 82 Å/min without PAM to 12 Å/min with 0.06 wt% of PAM. It is considered that this decrease in the removal rate of the PETEOS film as a function of the concentration of the PETEOS film is due to the formation of the polymer-coated layer on the PETEOS film caused by the high affinity of PAM to the PETEOS film. From the adsorption isotherms shown in Figure 6.13, the PETEOS film has high affinity to PAM, which induces PAM to adsorb on the surface of the PETEOS film. The PAM layer formed on the PETEOS layer reduces the possibility of the penetration of the abrasive particles to the PETEOS film and also decrease the friction force between abrasive particles and the PETEOS film. On the other hand, the removal rate for poly Si film remained at 2190 Å/min without PAM and 2178 Å/min with 0.06 wt% of PAM. Compared with the decrease of the removal rate for the PETEOS film, the change of the removal rate for poly Si film is small enough to ignore. It is expected that PAM scarcely adsorbs on poly Si film due to the low affinity between them as shown in Figure 6.13. Therefore, the PAM layer on poly Si is not thick enough to prevent the penetration of the abrasive particles to poly Si film even though the concentration of PAM is increased. Therefore, it is expected that poly Si film is directly exposed to the mechanical polishing during the poly isolation CMP process and it results in the small change of the removal rate of poly Si film with addition of PAM. As a result, the difference in the removal mechanism of each film achieves the drastic improvement in poly-Si-to-oxide selectivity values from 26.7 to 181.8 as the concentration of PAM increases from 0 to 0.06 wt%. Consequently, the formation of a polymer-coated layer by the selective adsorption of PAM on the PETEOS film plays a dominant role to improve poly-Si-to-oxide selectivity in poly isolation CMP. One can conclude that the control of selective adsorption of polymer regarding the characteristics of target films is a key technique to achieve poly-Si-to-oxide selectivity in the poly isolation CMP process.

6.3 Novel CMP for New Memory

Recently, flash memory—used in mobile phones, MP3s, digital cameras, and USBs—is a non-volatile memory device that solved a weak point of DRAM device of volatile movement. NAND flash memory in business today is possible up to 2 gigabyte in integrated diagram. It has a characteristic of action at high supply voltage of 10 to 15V.

Flash memory device applies threshold voltage transition of transistor with memory motion theory to floating gate made by polysilicon through the accumulating or discharge of electric charge. With the interaction of flash memory device structures, unbalanced polysilicon grain-size distribution increases threshold voltage distribution. The movement above 10 V voltage generated from devices is increasing. The technical limit of integration is known as 16 gigabit level. Therefore, there is a need for the next-generation non-volatile memory devices that are several integrations over 64 gigabit with tens of nanosecond write/erase time (Figure 6.16). Research has progressed in the areas of phase-change random access memory (PRAM), nano-floating gate memory (NFGM), polymer random access memory (PoRAM), and so on (Figure 6.17).

6.3.1 GST CMP for PRAM

Phase-change random access memory (PRAM) has been intensively studied as one of the candidates of a non-volatile memory to challenge conventional memories such as DRAM and flash memory, due to its fast

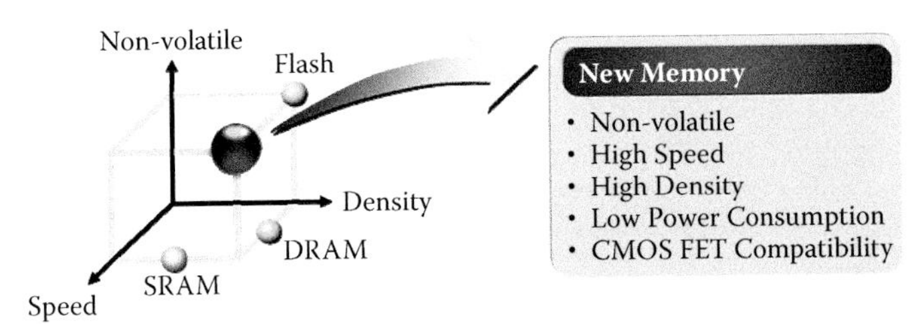

FIGURE 6.16 (See color insert) Comparison of new memory and conventional memory.

	Baseline 2004 Technologies		PRAM	NFGM	PoRAM	ReRAM
Storage Mechanism						
First Availability	2004	2004	~2006	>2006	>2010	~2010
Device Types	DRAM	NOR Flash	OUM	Engineered Tunnel Barrier or Nanocrystal	Bistable Switch	MIM
Cell Elements	1T1C	1T	1T1R	1T	1T	1T1R
F Value	90 nm	90 nm	100 nm	80 nm	45 nm	65 nm
Cell Size	$8F^2$ 0.065 μm^2	$12.5F^2$ 0.101 μm^2	$\sim 6F^2$ 0.06 μm^2	$\sim 6F^2$ 0.038 μm^2	$4F^2$ ~0.008 μm^2[1]	$6F^2$ 0.025 μm^2[C]
Write/Erase Time	50 ns	1us/1ms	<100 ns	10 ns	<10 ns	50ns/<100ns
Retention Time	64 ms	10–20 years	>10 years	>10 years	>1 year	>1 year
E/W Cycles	Infinite	>1E5	>1E13	>1E6	>1E15	>1E3

FIGURE 6.17 Classification of memory.

switching speed, good endurance, and compatibility with CMOS logic process. PRAM stores data according to the resistance change of chalcogenide (GeSbTe, GST) materials that is basically caused by the phase change of chalcogenide materials, and the phase change is followed by heat flux from the bottom electrode (BE) to contact dimension. Chalcogenide materials exhibit at least two states. The states are the amorphous and crystalline states, and transitions between these states may be selectively initiated. The amorphous state generally exhibits higher resistivity than the crystalline state. The phase change may be induced reversibly. Therefore, the memory may change from the amorphous to the crystalline state and may revert back to the amorphous state thereafter in response to temperature changes. In effect, each memory cell may be thought of as a programmable resistor, which reversibly changes between higher and lower resistance states. The phase change may be induced by resistive heating. As shown in Figure 6.18, PRAM manufacturing applies the CMP

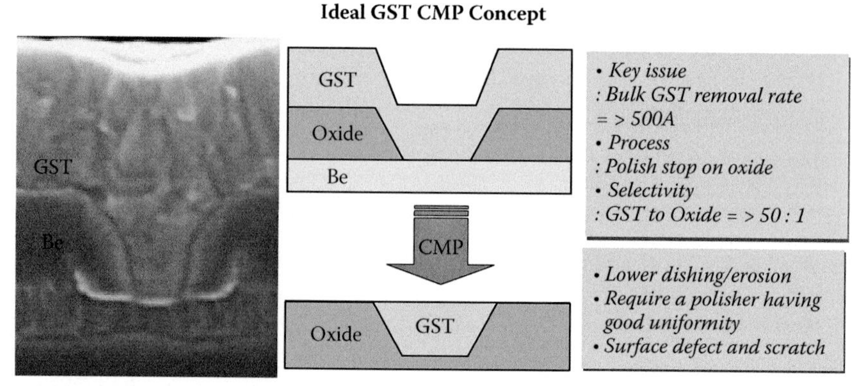

FIGURE 6.18 Schematic drawing of GST CMP process for on-axis confined structure in PRAM device.

technique to planarize the GST layers deposited on the front surfaces of wafers. In particular, to scale down PRAM beyond 256 Mbit (design rule less than 100 nm), achieving the writing current level of several hundred *u*A is an essential parameter. Therefore, GST was on-axis confined within a small pore, which may result in high planarity of GST surface. The SiO_2 film acts to isolate the adjacent cell with GST materials, and as a barrier that stops the polishing process immediately after complete removal of the GST materials. Besides, in the fabrication of PRAM beyond 256 Mb, the ring-shaped contact structure for small BEC formation is very important to process technologies. The writing current flows through the perimeter of the ring-shaped contact instead of the whole body of the contact. Obviously, the effective contact area of the ring-shaped contact has less dependency on the contact diameter because it is linearly proportional to the diameter of the defined contact. Thus the ring-shaped contact structure has robustness against the contact size variations. The CMP slurry needs to perform a high selectivity (>50:1) of polishing rate between GST and SiO_2 films. CMP slurry is composed of colloidal silica abrasive, surfactant, organic chemical, alkaline agent, titrant, and deionized water to control the removal selectivity of GST-to-SiO_2 films and TiN-to-SiO_2.

For CMP evaluation, 8-inch silicon wafers with a multilevel structure of NGST/SiO_2/Si were used. The as-deposited NGST film had an amorphous structure. The NGST film was deposited using a metal organic precursor under a nitrogen atmosphere at 350°C, giving a composition of approximately 25:23:52 (Ge:Sb:Te) by atomic percentage, as shown by the cross-sectional TEM energy-dispersive x-ray spectroscopy (EDS) analysis results listed in Figure 6.19. The bottom oxide film was deposited by the PETEOS method at 720°C. The thicknesses of the as-deposited NGST and oxide films were 200 and 100 nm, respectively. The films were polished on a CMP system (6EC, Strasbaugh, USA.) with a single polishing head and a polishing platen. We used an industry-standard CMP polishing pad (IC1000/Suba IV, Rohm and Haas Electronic Materials, USA.). The thickness variation of the NGST films on the wafers before and after CMP was

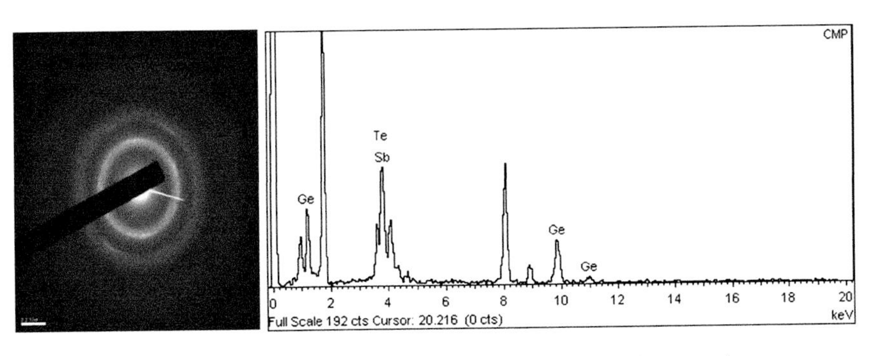

FIGURE 6.19 TEM energy-dispersive x-ray spectroscopy (EDS) analysis results.

measured with a spectroscopic phase-modulated ellipsometer (UVISEL, HORIBA Jobin Yvon, Japan). The oxide film thickness before and after CMP was estimated with a tabletop film analysis system (Nanospec 180, Nanometrics, USA). The contact angle was measured with a contact angle meter (DIGIDROP, GBX, France). The chemical bonding characteristics of the NGST film surface after dipping in the slurry were characterized by x-ray photoelectron spectroscopy (XPS) using an ESCA 2000 (V.G. Microtech, UK) system with a monochromatic Al Kα x-ray monochromatic source (linewidths = 0.85 eV, energy = 1487 eV).

Figure 6.20a shows the polishing rate of NGST film as a function of the tetramethylammonium hydroxide (TMAH) concentration. Without TMAH in the slurry, the polishing rate was 6 nm/min. With the TMAH addition, however, the NGST polishing rate drastically increased up to 242 nm/min at a TMAH concentration of 0.12 wt%. This striking difference in the polishing rates with and without TMAH resulted from the chemical reactions between the NGST film and the TMAH, which we explain later in detail. Beyond a TMAH concentration of 0.12 wt%, however, the polishing rate of NGST slightly decreased. This behavior was related to the way that TMAH effectively influences chemical reactions at the NGST film surface, such as hydrophobic or hydrophilic interactions. The characteristics of the reaction of TMAH with the NGST film surface can be estimated from the surface tension by measuring the contact angle. In general, wettability can be quantitatively evaluated in terms of the spreading coefficient, which is the energy difference between the solid substrate and the contacting liquid phase. The interaction of the interfacial tensions at the liquid–vapor–solid junction is described by the Young equation as follows:

$$\cos\theta = \frac{r_{SV} - r_{SL}}{r_{LV}} \tag{6.1}$$

where θ is contact angle ($\theta > 0°$), and r_{SV}, r_{SL}, and r_{LV} are the effective interfacial tensions.

After dipping the NGST film in the slurry for 1 min at 45°C, the contact angle of the NGST film surface was measured as a function of the TMAH concentration, as shown in Figure 6.20b. The contact angle decreased drastically with increasing TMAH concentration up to 0.12 wt% and then gradually increased. The contact angle results shown in Figure 6.20b coincided with the inverse of the NGST film polishing rate curve in Figure 6.20a. The hydrophilicity indicated by a small contact angle means that adhesion of TMAH molecules on the NGST film surface, more so than cohesion, plays a dominant role in enhancing the impact probability of molecules for the chemical reaction to etch the oxidized surface of the NGST film. The enhanced impact probability results in an accelerating chemical and mechanical reaction speed between the colloidal silica abrasives and

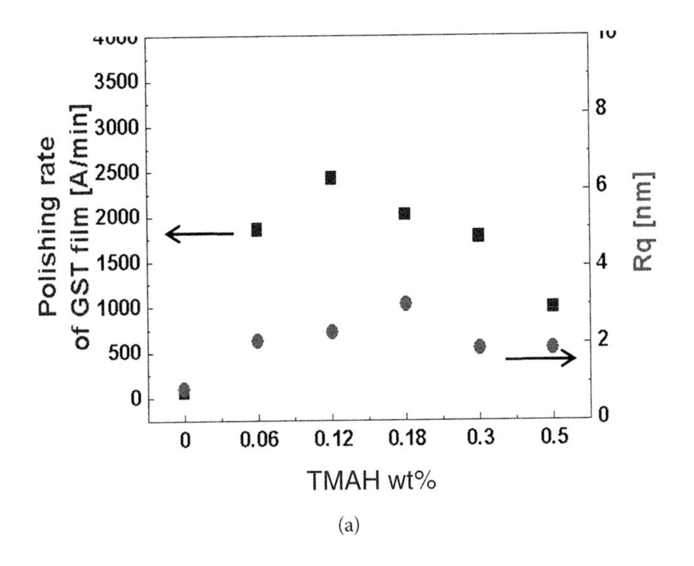

(a)

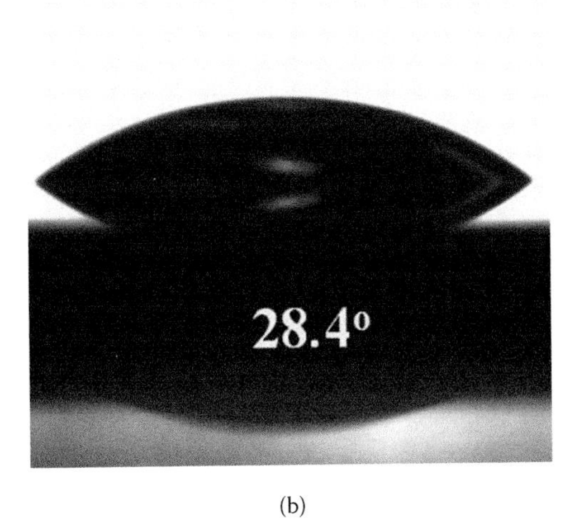

(b)

FIGURE 6.20 (a) Polishing rate of NGST film as a function of the TMAH concentration, (b) contact angle of the NGST film surface.

the NGST film surface during CMP. As a result, a slurry with enhanced hydrophilicity initially produces a higher polishing rate. Note that TMAH molecules initially etch off the oxidized surface of the as-deposited NGST film, and the exposed, unoxidized NGST surface is then chemically oxidized, resulting in a hydrophilic surface. The hydrophilicity of the NGST film surface, however, becomes progressively lower once the TMAH concentration exceeds a specific value (0~0.12 wt% in this experiment). This is probably associated with the formation of a polymer layer enhanced by excessive TMAH molecules on the NGST film surface, making the surface less hydrophilic and suppressing the direct mechanical contact between the colloidal silica abrasives and the unoxidized NGST film surface. As a result, the polishing rate of NGST film decreases with increasing contact angle. These results for NGST film CMP—namely, the relation between the polishing rate and contact angle—are the same as those that we previously reported for polysilicon CMP. Therefore, this study has provided key information on using TMAH in colloidal silica slurry to obtain a high polishing rate of NGST film during CMP.

We conducted an additional dipping test to confirm the validity of the contact-angle mechanism of NGST film CMP. The as-deposited NGST film was dipped into the slurry for one minute at 45°C. After dipping, the chemical composition of the NGST film surface was characterized by XPS, which is one of the best ways to examine the chemical binding characteristics of film surfaces and further explore the chemical reaction between TMAH and the NGST surface. Figure 6.21a shows the XPS spectra for Ge 2p. Before dipping, there was a strong GeO_2 peak at 1219.8 eV; after dipping, this peak was weakened. Thus, TMAH is a strong etchant of GeO_2 on the NGST film surface. Figure 6.21b shows the XPS spectra for Sb 3d. The Sb $3d^{3/2}$ peaks for Sb_2O_5 and metallic Sb bonding occurred at 539.8 eV and 537–538 eV, respectively. The Sb $3d^{5/2}$ peaks for Sb_2O_5 and metallic Sb bonding occurred at 530.4 eV and 529–530 eV, respectively. Here, Sb metallic bonding refers to Sb–Te or Sb–Ge bonds and the formation of Sb clusters in the film. Before dipping, the peak position for Sb was between the Sb homopolar and metallic bonds, so there could be equivalent numbers of Sb homopolar and Sb–Te or Sb–Ge bonds in the amorphous NGST film. After dipping, the intensities of the Sb $3d^{3/2}$ and Sb $3d^{5/2}$ peaks for Sb_2O_5 decreased significantly. This indicates that Sb_2O_5 was easily etched by the TMAH-based slurry. In the case of tellurium oxide, the Te $3d^{5/2}$ peaks for Te metallic bonding and Te oxide bonding occurred at 572.5–574 eV and 576–577 eV, respectively. The Te $3d^{3/2}$ peaks for Te metallic bonding and Te oxide bonding occurred at 583–584 eV and 586–587 eV, respectively. As shown in Figure 6.21c, there was no significant difference in the XPS spectra for the Te 3d peaks before and after dipping. This indicates that Te oxide bonding is stronger than the oxide bonding of Ge and Sb.

The results suggest a possible CMP mechanism for NGST film. The TMAH initially etches off the oxidized film surface (particularly GeO_2 and Sb_2O_5), and the colloidal silica abrasives then directly contact the unoxidized NGST film surface to perform mechanical polishing. In addition, hardness measurement of the NGST film showed that its hardness (3.2 GPa) was approximately three times smaller than that of SiO_2 film (9 GPa). As a result, the colloidal silica abrasives in the TMAH-based slurry significantly enhance the polishing rate of NGST film by etching off the oxidized surface and applying direct mechanical polishing to the soft NGST film surface.

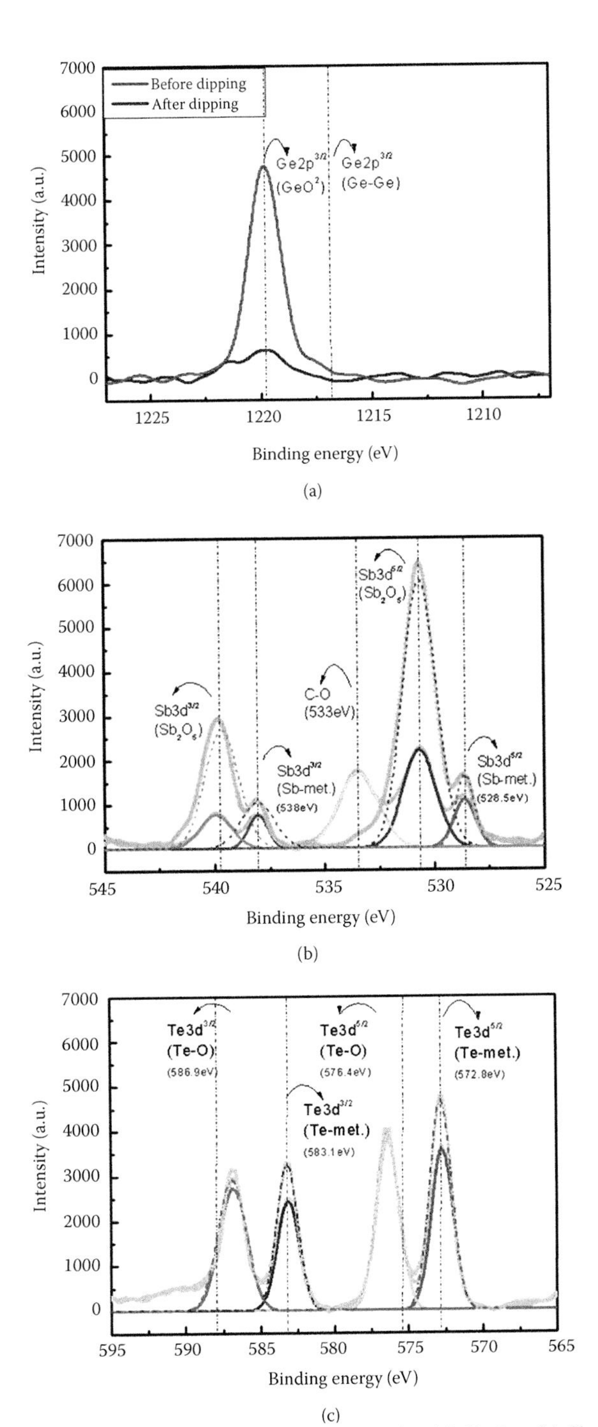

FIGURE 6.21 The XPS spectra of NGST film for (a) Ge 2p, (b) Sb 3d, and (c) Te 3d.

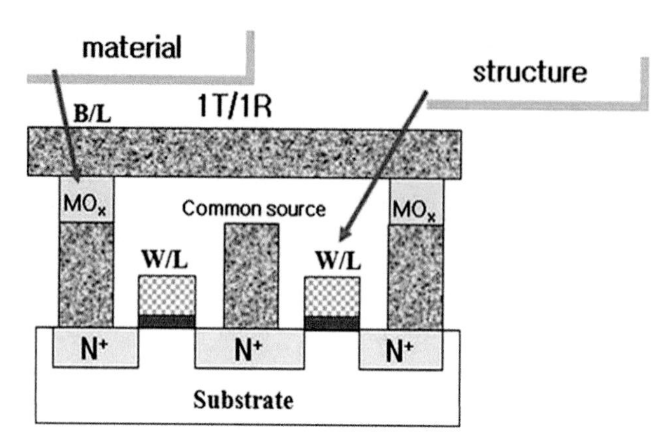

FIGURE 6.22 The structure of ReRAM.

6.3.2 Novel CMP for ReRAM

ReRAM can produce a device by 1T/1R or 1R structure (Figure 6.22). In the case of producing by 1T1R structure, after deposition of a resistance change material except for CMP for STI, the CMP process is considered. ReRam is not sure about its effect regarding contact-area dimension like PRAM. If it is not produced by extremely confined structure, the use of CMP process depends on the processing choice.

Recently, in the new ReRAM non-volatile memory, a nickel oxide film with titanium (Ti:NiO) was doped to increase voltage for memory erasures. Operations required only 5 ns, about 10,000 times faster than before, and with resistance fluctuations reduced to 1/10 that of conventional ReRAMs. Optimizing the voltage applied to a transistor reduced the current need to erase memory to $\leq$100 µA.

The prototype ReRAM device offers low fluctuation of resistance value even during high-speed operation. The ReRAM technology "is amenable to miniaturization and can be manufactured inexpensively," so it is seen as an alternative to flash, the company explained. "If further minute non-volatile memory can be realized using ReRAM, there is potential for higher performance of mobile devices."

References

Ali, I., et al., *Solid-State Technol.*, October 1994.

America, W. G., and S. V. Babu, *Electrochem. Solid-Atate Lett.*, 7, G327, 2004.

Baraban, P., V. V. Bulavinov, and P. P. Konorov, *Electronics of SiO_2 Layers on Silicon*, Izd. LGU, Leningrad, 1988.

Basim, G. B., and B. M. Boudgil, *J. Colloid Interface Sci.*, 256, 137, 2002.

Beecroft, L., and C. K. Ober, *Chem. Mater.*, 9, 1302, 1997.

Billmeyer, F. W., *Textbook of Polymer Science*, Wiley Interscience, New York, 1984, p. 7.

Bohmer, M. R., O. A. Evers, and J. M. H. M. Scheutjens, *Macromolecules*, 23, 2288, 1990.

Boning, D., *MRS Bull. B.*, 27, 761, 2002.

Boning, D., and B. Lee, *MRS Bull.*, 27, 761, 2002.

Boning, D., B. Lee, W. Baylies, N. Poduje, P. Hester, J. Valley, and C. Koliopoulos, International CMP Symposium, Tokyo, 2000.

Boning, D., B. Lee, C. Oji, D. Ouma, T. Park, T. Smith, and T. Tugbawa, Pattern dependent modeling for CMP optimization and control, *Proc. Symposium P: Chemical Mechanical Polishing* (*MRS Spring Meeting*), San Francisco, CA, 1999, p. 761.

Boning, D. S., and O. Ouma, Chemical mechanical polishing in silicon processing, in *Semiconductors and Semimetals*, Vol. 63, Academic Press, 2000, chap. 4, p. 108.

Bonner, B. A., A. Iyer, D. Kumar, T. H. Osterheld, A. S. Nickles, and D. Flynn, Development of a direct polish process for shallow trench isolation modules, *CMP-MIC Spring Meeting, Proc. Symposium: Chemical Mechanical Planarization for ULSI Multilevel Interconnection*, 2001, p. 572.

Boyd, J. M., and J. P. Ellul, *J. Electrochem. Soc.*, 143, 3718, 1996.

Burke, P., *VMIC*, pp. 379–384, 1991.

Cardinaud, C. H., G. Lemperiere, M. C. Peignon, and P. Y. Jouan, *App. Surf. Sci.*, 68, 595, 1993.

Carter, P. W., and T. P. Johns, *Electrochem. Solid-State Lett.*, 8, G221, 2005.

Chang, E., et al., *IEDM Tech. Digest*, pp. 499–502, 1995.

Chekina, G., and L. M. Keer, *J. Electrochem. Soc.*, 145, 2100, 1998.

Cheng, J. Y., T. F. Lei, T. S. Chao, D. L. W. Yen, B. J. Jin, and C. J. Lin, *J. Electrochem. Soc.*, 144, 315, 1997.

Cheng, J. Y., T. F. Lei, and T. S. Chao, *Jpn. J. Appl. Phys.*, 36, 1319, 1997.

Cho, C. W., S. K. Kim, U. Paik, J. G. Park, and W. M. Sigmund, *J. Mater. Res.*, 21, 473, 2006.

Choibowski, S., and M. Wisniewska, *Colloids Surf. A*, 208, 131, 2002.

Clark, M. G., *Proc. of IEE Circuits Devices Syst.*, 141, 3, 1994.

Clear, S. C., and P. F. Nealey, *J. Colloid Interf. Sci.*, 213, 238, 1999.

Contolini, R. J., Bernhardt, A. F, and Mayer, S. T. *J. Electrochem. Soc.*, 141, 2503, 1994.

Cook, L. M., *J. Non-Cryst. Solids*, 120, 152, 1990.

Cooperman, S. S., A. I. Nasr, and G. J. Grula, *J. Electrochem. Soc.*, 142, 3180, 1995.

Coppetta, J., C. Rogers, A. Phillipossian, and F. Kaufman, *Proc. 2nd CMP-MIC*, 307, 1997.
Cross, L. E., *Ferroelectrics*, 76, 241, 1987.
De Angelis, C., Rizzo, S. Contarini, and S. P. Howlett, *App. Surf. Sci.*, 51, 177, 1991.
Depasse, J., and A. Watillon, *J. Colloid and Interface. Sci.*, 33, 430, 1970.
Divecha, R., et al., *Proc. of the 1st IEEE International Workshop on Statistical Metrology*, June 1996.
Djuricic, B., S. Pickering, *J. Eur. Ceram. Soc.*, 19, 1925, 1999.
Fierro, J. L. G., S. Mendioroz, and A. M. Olivan, *J. Colloid Interface Sci.*, 100, 303, 1984.
Fukuda, T., Y. Shimizu, M. Yoshise, M. Hashimoto, and T. Kumagai, *Proc. 3rd Int. Symp. Advanced Science and Technology of Silicon Materials, The Japan Society for the Promotion of Science*, 2000, p. 382, Kona, Hawaii.
Gorantla, V. R. K., Babel, A., Pandija, S., and Babu, S. V. *Electrochem. Solid-State Lett.*, 8, G131, 2005.
Hackley, V. A., *J. Am. Ceram. Soc.*, 80, 2315, 1997.
Hackley, V. A., and S. G. Malghan, *J. Mater. Sci.*, 29, 4420, 1994.
Healy, T. W., in *The Colloid Chemistry of Silica*, H. E. Bergna, Ed., American Chemical Society, Washington, DC, 1994, p. 147.
Hernandez, J., P. Wrschka, Y. Hsu, T.-S. Kuan, G. S. Oehrlein, H. J. Sun, D. A. Hansen, J. King, and M. A. Fury, *J. Electochem. Soc.*, 146(12), 4647, 1999.
Hetherington, D., et al., *CMP-MIC*, pp. 74–81, 1996.
Hiemenz, P. C., and R. Rajagopalan, *Principles of Colloid and Surface Chemistry*, Marcel Dekker, New York, 1997, chap. 13, pp. 585–592.
Hirai, K., H. Ohtsuki, T. Ashizawa, and Y. Kurata, High performance CMP slurry for STI, Hitachi Chemical Tech. Report No. 35, 17, 2000.
Hisamune, Y. S., K. Kanamori, T. Kubota, Y. Suzuki, M. Tsukiji, E. Hasegawa, A. Ishitani, and T. Okazawa, *IEDM Tech. Dig.*, 19, 1993.
Homma, Y., T. Furusawa, H. Morishima, and H. Sato, *Solid-State Electron.*, 41, 1005, 1997.
Horiike, Y., H. Sakaue, H. Shindo, and S. Shingubara, *JKPS*, 26, S75, 1993.
Horn, R. G., *J. Am. Ceram. Soc.*, 73, 1117, 1990.
Hoshino, T., Y. Kurata, Y. Terasaki, and K. Susa, *J. Non-Cryst. Solids*, 283, 129, 2001.
Hu, Y. Z., R. J. Gutmann, and T. P. Chow, *J. Electrochem. Soc.*, 145, 3919, 1998.
Hunter, R. J., *Introduction to Modern Colloid Science*, Oxford University Press, New York, 1993, chap. 1, p. 204.
Hussein, G. A. M., *J. Anal. Appl. Pyrolysis*, 37, 111, 1996.
Iler, R. K., *The Chemistry of Silica*, John Wiley & Sons, New York, 1979.
Itoh, A., M. Imai, and Y. Arimoto, *Jpn. J. Appl. Phys.*, 37, 1697, 1998.
Israelachvili, J. N., *Intermolecular and Surface Forces*, Academic Press, San Diego, CA, 1987.
Kang, H. G., T. Katoh, W. M. Lee, U. Paik, and J. G. Park, *Jpn. J. App. Phys.*, 43, L1, 2004.
Kang, H. G., T. Katoh, H. S. Park, U. Paik, and J. G. Park, *Jpn. J. App. Phys.*, 43, L365, 2004.
Kang, H. G., T. Katoh, H. S. Park, U. Paik, and J. G. Park, *Jpn. J. App. Phys.*, 44, 4752, 2006.
Katoh, T., H. G. Kang, U. Paik, and J. G. Park, *Jpn. J. App. Phys.*, 42, 1150, 2003.
Katoh, T., M. S. Kim, U. Paik, and J. G. Park, *Jpn. J. Appl. Phys.*, 43, L217, 2004.
Katoh, T., S. J. Kim, U. Paik, and J. G. Park, *Jpn. J. Appl. Phys.*, 42, 5430, 2003.

Katoh, T., B. G. Ko, J. H. Park, H. C. Yoo, J. G. Park, and U. G. Paik, *J. Korean Phys. Soc.*, 40, 180, 2002.
Katoh, T., J. G. Park, W. M. Lee, H. Jeon, U. Paik, and H. Suga, *Jpn. J. Appl. Phys.*, 41, L443, 2002.
Kaufman, F. B., D. B. Thomson, R. E. Broadie, M. A. Jaso, W. L. Guthriem, D. J. Pearson, and M. B. Small, *J. Electrochem. Soc.*, 138(11), 3460, 1991.
Kim, D. C., W. C. Shin, J. D. Lee, J. H. Shin, J. H. Lee, S. H. Hur, I. H. Baik, et al., *IEDM Tech. Dig.*, 919, 2002.
Kim, D. H., H. G. Kang, S. K. Kim, U. Paik, and J. G. Park, *Jpn. J. App. Phys.*, 45, 4893, 2006.
Kim, J. P., Y. S. Jung, J. G. Yeo, U. Paik, J. G. Park, and V. A. Hackley, *J. Kor. Phys. Soc.*, 39, 197, 2001.
Kim, J. P., U. Paik, Y. G. Jung, T. Katoh, and J.-G. Park, *Jpn. J. Appl. Phys.*, 41, 4509, 2002.
Kim, J. Y., S. K. Kim, U. Paik, T. Katoh, and J. G. Park, *J. Korean Phys. Soc.*, 41, 413, 2002.
Kim, K., *IEDM Tech. Dig.*, 13.5, 2005.
Kim, S. D., I. S. Hwang, and H. M. Park, *J. Vac. Sci. Technol. B*, 20, 918, 2002.
Kim, S. K., S. Lee, U. Paik, T. Katoh, and J. G. Park, *J. Mater. Res.*, 18, 2163, 2003.
Kim, S. K., S. Lee, U. Paik, T. Katoh, and J. G. Park, *Jpn. J. App. Phys.*, 43, 7427, 2004.
Kim, S. K., U. Paik, S. G. Oh, T. Katoh, and J. G. Park, *Jpn. J. Appl. Phys.*, 42, 1227, 2003.
Kim, T., J. Kim, and J. Om, *JKPS*, 35, S861, 1999.
Koh, C.-G., I.-S. Yeo, S.-H. Pyi, and S.-K. Lee, *JKPS*, 35, S1038, 1999.
Kubo, N., in *Handotai heitanka CMP gijyutsu* (CMP technology for semiconductor planarization), T. Doi, T. Kasai, T. Nakagawa, Eds., Kougyo-chosakai, Tokyo, 1998, p. 124.
Laparra, O., and M. Weling, *Proc. Int. Symp. Chemical Mechanical Planarization II, San Diego, 1998*, Electrochemical Society, Pennington, 1998, p. 98.
Lee, B., *Impact of Nanotopography on STI CMP in Future Technologies*, PhD thesis, Dept. of Electrical Engineering and Computer Science, Massachusetts Institute of Technology, Cambridge, Massachusetts, 2002.
Lee, B., D. Boning, W. Baylies, N. Poduje, P. Hester, Y. Xia, J. Valley, et al., Materials Research Society (MRS) Spring Meeting, San Francisco, CA, April 2001.
Lee, B., T. Gan, D. Boning, P. Hester, N. Poduje, and W. Baylies, in *Advanced Semiconductor Manufacturing Conference*, Boston, MA, Sept. 2000, p. 425.
Lee, J. D., S. H. Hur, and J. D. Choi, *IEEE Electron Device Lett.*, 23, 5, 2002.
Lee, L. T., and P. Somasundaran, *Langmuir*, 5, 854, 1989.
Lee, S. H., Z. Lu, S.V. Babu, and E. Matijevi, *J. Mater. Res.*, 17, 2744, 2002.
Lenzlinger, M., and E. H. Snow, *J. Appl. Phys.*, 40, 278, 1969.
Lewis, J. A., *J. Am. Ceram. Soc.*, 83, 2341, 2000.
Li, S. H., and R. O. Miller, *Chemical Mechanical Polishing in Silicon Processing*, Academic Press, New York, 2000, p. 142.
Lin, C. F., W. T. Tseng, and M. S. Feng, *J. Electrochem. Soc.*, 146(5), 1984, 1999.
Lin, C. F., W. T. Tseng, M. S. Feng, and Y. L. Wang, *Thin Solid Films*, 347, 248, 1999.
Luo, J., and D. A. Dornfeld, *IEEE Trans. Semiconductor Manufacturing*, 16, 469, 2003.
Maca, K., M. Trunec, and J. Cihlar, *Ceram. Int.*, 28, 337, 2002.
Mahanty, J., and B.W. Ninham, *Dispersion Forces*, Academic Press, New York, 1979.
McCormick, P. G., T. Tsuzuki, J. S. Robinson, and J. Ding, *Adv. Mater.*, 13, 1008, 2001.
Milonijic, S. K., *Colloids Surf.*, 63, 113, 1992.

Muller, T., R. Kumpe, H. A. Gerber, R. Schmolke, F. Passek, and P. Wagner, *Microelectronic Eng.*, 56, 123, 2001.
Nojo, H., M. Kodera, and K. Nakata, *IEEE*, 96, 349, 1996.
Otsuga, K., H. Kurata, S. Node, Y. Sasago, T. Arigane, T. Kawamura, and T. Kobayashi, *IEICE Trans. Electron.*, E90-C 4, 2007.
Ouma, D., C. Oji, D. Boning and J. Chung, *Proc. CMP-MIC Conf. IMIC*, 1998, p. 20.
Oya, A., F. Beguin, K. Fujita, and R. Benoit, *J. Mater. Sci.*, 31, 4609, 1996.
Paik, U., V. A. Hackley, S. C. Choi, and Y. G. Jung, *Coll. Surf.*, A135, 77, 1998.
Paik, U., V. A. Hackley, J. Lee, and S. Lee, *J. Mater. Res.*, 18, 1266, 2005.
Paik, U., V. A. Hackley, and H. W. Lee, *J. Am. Ceram. Soc.*, 82, 833, 1999.
Paik, U., J. P. Kim, T. W. Lee, Y. G. Jung, J. G. Park, and V. A. Hackley, *J. Kor. Phys. Soc.*, 39, 201, 2001.
Palla, B. J., and D. O. Shah, *IEEE/CPMT Intl. Electronics Manufacturing Technology Symposium*, 1999.
Palla, B. J., and D. O. Shah, *J. Coll. Inter. Sci.*, 223, 102, 2000.
Park, H. S., K. B. Kim, C. K. Hong, U. I. Chung, and M. Y. Lee, *Jpn. J. Appl. Phys.*, 37, 5849, 1998.
Park, J. G., T. Katoh, W. M. Lee, H. Jeon, and U. Paik, *Jpn. J. Appl. Phys.*, 42, 5420, 2003.
Park, J. G., T. Katoh, H. C. Yoo, D. H. Lee, and U. Paik, *Jpn. J. App. Phys.*, 41, L17, 2002.
Park, J. G., T. Katoh, H. C. Yoo, and U. G. Paik, *Proc. 5th Int. Symp. Chemical Mechanical Polishing* (201st Meeting of the Electrochemical Society, Philadelphia, PA), 2002, p. 202.
Park, J. G., T. Katoh, H. C. Yoo and J. H. Park, *Jpn. J. Appl. Phys.*, 40, L857, 2001.
Park, J. G., T. Katoh, H. C. Yoo, and J. H. Park, *Jpn. J. Appl. Phys.*, 40, L857, 2001.
Park, J. H., S. H. Hur, J. H. Lee, J. T. Park, J. S. Sel, J. W. Kim, S. B. Song, et al., *IEDM Tech. Dig.*, 873, 2004.
Pavan, P., R. Bez, P. Olivo, and E. Zanoni, *Proc. of IEEE*, 85, 1248, 1997.
Philipossian, A., and M. Hanazono, Tribology and fluid dynamics characterization of cerium oxide slurries, www.innovative-planarization.com, 2001.
Preston, F. W., *J. Soc. Glass Technol.*, 120, 214, 1927.
Pye, J. T., H. W. Fry, and W. J. Schaffer, *Solid State Tech.*, December, 65, 1995.
Quirk, M., and J. Serda, *Semiconductor Manufacturing Technology*, Prentice Hall, New Jersey, 2001, chap. 9, p. 199.
Raghavan, S. R., H. J. Walls, and S. A. Khan, *Langmuir*, 16, 7920, 2000.
Rajan, K., R. Singh, J. Adler, U. Mahajan, Y. Rabinovich, B. Moudgil, *Thin Solid Films*, 308-309, 529, 1997.
Ravi, K. V., Wafer flatness requirements for future technology, *Future Fab International*, Summer 1999.
Reed, J. S., *Principles of Ceramics Processing*, 2nd ed., Wiley Interscience, New York, 1995, chap. 17, p. 323.
Rozman, M., and M. Drofenik, *J. Am. Ceram. Soc.*, 81, 1757, 1998.
Russel, W. B., *J. Rheol.*, 24, 287, 1980.
Scheutjens, J. M., and G. J. Fleer, *J. Phys. Chem.*, 83, 1619, 1979.
SEMI Document M43: Guide for Reporting Wafer Nanotopography, 2001.
SEMI Draft Document 3089: A Guide for Reporting Wafer Nanotopography, 1999.
Semiconductor Industry Association. *Front end processes in international technology roadmap for semiconductor*, p. 4.
Semiconductor Industry Association. *Process integration, devices, and structures in international technology roadmap for semiconductor*, p. 23, 2006.

Seo, Y. J., and W. S. Lee, *J. Korean Phys. Soc.*, 48, 1651, 2006.
Seo, Y. J., S. W. Park, C. B. Kim, S. Y. Kim, W. S. Lee and J. S. Park, *J. Korean Phys. Soc.*, 42, 421, 2003.
Shen, Q., D. Mu, L. W. Yu, and L. Chen, *J. Col. Inter. Sci.*, 275, 30, 2004.
Singh, R. K., S.-M. Lee, K.-S. Choi, G. B. Basim, W. Choi, Z. Chen, and B. M. Moudgil, *J. Mater. Res. Bull.*, 27, 752, 2002.
Sivaram, S., H. Bath, R. Legegett, A. Maury, K. Monning, and R. Tolles, *Solid State Tech.*, 35-5, 87, 1992.
Sivaram, S., H. Bath, R. Legegett, A. Maury, K. Monning, and R. Tolles, *Solid State Tech*, 97, May 1992.
Sjöberg, S., *J. Non-crystalline Solids*, 196, 51, 1996.
Smekaiin, K., *Solid State Tech.*, 40, 187, 1997.
Steigerwald, J. M., S. P. Murarka, and R. J. Gutmann, *Chemical Mechanical Planarization of Microelectronic Materials*, Wiley Interscience, New Jersey, 1997, p. 141.
Steigerwald, J. M., R. Zirpoli, S. P. Muraka, and R. J. Gutmann, *J. Electrochem. Soc.*, 142(10), 2841, Oct. 1994.
Stine, B., D. S. Boning, J. E. Chung, *IEEE Trans. Semiconductor Manuf.*, 10, 24, 1997.
Stine, B., D. Ouma, R. Divecha, D. Boning, J. Chung, D. Hetherrington, I. Ali, et al., A closed-form analytic model for ILD thickness variation in CMP processes, *Proc. CMP-MIC Conf.*, Santa Clara, CA, Feb. 1997.
Sze, S. M., and K. Ng, *Physics of Semiconductor Devices*, 3rd ed., Wiley Interscience, Hoboken, NJ, 2007.
Tamura, N., Niwa, H., Hatanaka, M., Kase, M., and Fukuda, T. *197t h ECS Meeting, Toronto, Ontario, Canada*, May, 2000.
Togrul, H., and N. Arslan, *Carbohydrate Polymers*, 54, 63, 2003.
Tsukruk, V. V., and V. N. Bliznyuk, *Langmuir*, 14, 446, 1998.
Tung, T.-L., A method for die-scale simulation for CMP planarization, *Proc. of SISPAD Conf.*, Cambridge, MA, Sept. 1997.
Vedula, R. R. and Spencer, H. G. *Colloids Surf.*, 58, 99, 1991.
Wagner, D., J. *Vac. Sci. Technol.*, 15, 518, 1978.
Wang, L., W. M. Sigmund, and F. Aldinger, *J. Am. Ceram. Soc.*, 83, 697, 2000.
Wolf, S., *Silicon Processing for the VLSI Era Process Integration*, Vol. 2, Lattice Press, Sunset Beach, CA, 1990, chap. 13, p. 24.
Wolf, S., *Silicon Processing for the VLSI Era*, Vol. 4, Lattice Press, Sunset Beach, CA, 2002, chap. 8, pp. 375–390.
Wrschka, P., J. Hernandez, G. S. Oehrlein, and J. King, *J. Electrochem. Soc.*, 147(2), 706, 2000.
Xia, B., I. W. Lenggoro, and K. Okuyama, *Chem. Mater.*, 14, 2623, 2002.
Xu, C. S., E. Zhao, R. Jairath, and W. Krusell, *J. Electrochem. Solid-State Lett.*, 1, 4, 1998.
Yang, J., S. Mei, and J. M. F. Ferreira, *J. Am. Ceram. Soc.*, 83, 1361, 2000.
Yim, Y. S., K. S. Shin, S. H. Hur, J. D. Lee, I. H. Baik, H. S. Kim, S. J. Chai, et al., *IEDM Tech. Dig.*, 34.1.1, 2003.
Yu, C., P. C. Fazan, V. K. Mathews, and T. T. Doan, *Appl. Phys. Lett.*, 61, 1344, 1992.
Zaman, A. A., B. M. Moudgil, A. L. Fricke, and H. El-Shall, *J. Rheol.*, 40, 1191, 1996.
Zhou, Z., P. J. Scales, and D. V. Boger, *Chem. Eng. Sci.*, 56, 1290, 2001.

Index

O

P